William Ellis

Ellis's Husbandry

Abridged and Methodized Comprehending the Most Useful Articles of Practical

Agriculture

William Ellis

Ellis's Husbandry
Abridged and Methodized Comprehending the Most Useful Articles of Practical Agriculture

ISBN/EAN: 9783741101854

Manufactured in Europe, USA, Canada, Australia, Japa

Cover: Foto ©berggeist007 / pixelio.de

Manufactured and distributed by brebook publishing software (www.brebook.com)

William Ellis

Ellis's Husbandry

ELLIS'S

HUSBANDRY,

ABRIDGED AND METHODIZED:

COMPREHENDING THE MOST USEFUL ARTICLES OF

PRACTICAL AGRICULTURE.

IN TWO VOLUMES.

VOL. I.

PREFACE,

BY THE EDITOR.

WILLIAM ELLIS, of *Little Gaddefden,* near *Hempstead* in *Hertfordshire,* lived near fifty years on one farm in that parish. His education was something, not much superior to that of the general run of common farmers, but he inherited from nature strong and active parts, which enabled him to rise into a sphere superior to his brethren, and make his name familiar throughout all Europe; for his works have been quoted and commended by numerous foreign writers on the subject of husbandry. This is not surprising, for although his faults were very numerous, yet being a plain farmer, dependant only on his skill in common husbandry for many years, his practice could not fail of being extensive, and his observations numerous; so that his works, as he

borrowed

borrowed nothing from others, were really original, and contained in numerous inſtances more genuine knowledge than far more ſhining performances abounded with. It may be aſked, how a man poſſeſſing a good underſtanding, long practice, and attentive obſervation, ſhould produce ſo many faults as *Ellis?* and in truth his faults are ſo numerous, that they have prejudiced many good judges againſt his works; inſomuch that ſome will not allow him any merit, and *others are very ſparing in their praiſes of him.* A ſhort enquiry into this part of his life will explain the ſeeming contradiction.

His firſt work, *The Timber Tree Improved,* was what one may call the ſpontaneous production of his pen; the ſucceſs of which work (for it ſaw three editions in leſs than three years) induced *Oſborne* the bookſeller to recommend other undertakings of the ſame ſort to him; then it was that he came to monthly works, and more voluminous productions; in which, to make up a ſtipulated quantity agreed upon with his bookſeller, he gave into all thoſe random ridiculous

PREFACE.

lous details which have so much disgraced his page. But these works, bad as some parts of them were, introduced him to much travelling in the character of a seedsman—a seller of implements—and in short, as a man who executed any sort of country business at a given price. Any person in Great-Britain might send for him, on paying for his time and expences. This part of his business, carried him over most parts of the kingdom, and wherever he went he was never backward in noting all that was peculiar in the conduct of their agriculture. Thus had he, besides long experience at home, the fortunate opportunity of viewing the husbandry of most parts of the kingdom.

With such united advantages, he certainly possessed a fund of country knowledge rarely known to be the share of one man: his latter works ought therefore to have been peculiarly valuable; but having engaged for larger quantities of MS. than his materials of real excellence would allow, all his pieces are nearly equal in being too much filled with trash.

This did his reputation much mischief, and at last injured him so much with the public, that he no longer found any pecuniary advantage in writing, but stuck to his farm; and very wisely depended on that alone. Another circumstance contributed not a little to damping his fame as a publick farmer. Many gentlemen fond of farming called on him at *Gaddesden* in expectation of seeing on his farm, every thing excellent in husbandry that he had ever mentioned; but instead of that, they sometimes found his farm in bad common order, owing to his frequent absence; as to tools, of which he wrote so much, few were to be seen, and those bad, and in every respect a very poor appearance of spirit or conduct. This did *Ellis*'s reputation no small mischief, though very unjustly; for nothing could be more absurd than expecting him to practise all he recommended; he depended on his farm for bread, and would therefore have been very imprudent to *practise* with as little ceremony as he *wrote*. He went on in a common stile, and in that was really a good common farmer on his own farm; but those who visited him expected to see the *most excellent*

excellent three pulley drill *plough*; and the *admirable fine horse break* at work in every field; and at least half the farm drilled and horse-hoed. Their disappointment only proves that *Ellis* had more sense than they imagined: and as to other varieties of instruments, he procured and sold them to any persons, but it did not therefore follow that he was to keep them ready made on an uncertainty of sale. Such gentlemen, therefore, as condemned *Ellis* as a writer because he had not to *shew* all on which he *wrote*, were at least unreasonable in their expectations. Truth and good sense, are undoubtedly such in an author's page, though not in his practice; of which many instances might be quoted from various parts of his works—instances which will ever prove his deep knowledge of husbandry, whether he practised them or not.

Suffice it to say, that he had strong parts—long experience—attentive observation—and numerous opportunities—I ask no further. You may tell me of ten thousand absurdities if you please, but certain I am that such

such a man muſt in his work, abound with juſt and practical ideas. And that this is really the caſe with *Ellis*, I hope the following pages will clearly aſcertain.

His writings have been much neglected for near thirty years; I think very unjuſtly: becauſe in works that concern the practice of any art, men ſhould willingly overlook abſurdities, in order to reap profit by the ſterling good ſenſe ſcattered among them: thoſe that will not take ſome trouble of this ſort, deſerve not to make great progreſs. The neglect into which they are fallen is certainly owing to the very numerous paſſages either abſurd, trivial, or tedious; all theſe paſſages we ought in candour to attribute to his connections with his bookſeller, who inſiſted on certain quantities of monthly matter; to make up which the author wrote not what was beſt, but what came firſt: his real and genuine knowledge which every where ſhines through even this cloud of rubbiſh, is fully ſufficient to ſhew how well he underſtood huſbandry, badly as he might ſometimes write about it.

PREFACE.

The design of this edition of his works, is to give the world all his observations that seem important, and practical; excluding all those passages which it is supposed he wrote more to form a bulky pamphlet, than for any value he thought they would be of to the public. In a word, to give such a book as it is presumed *Ellis* himself would have given, had his unbiassed judgment been singly his director. Upon this plan all his gypsies, wenches, thieves, rogues, &c. are discarded, and his old woman's tales which filled a page but diminished from its value, are thrown aside. Other deductions are made which however require an explanation.

Ellis made a traffick, sometimes profitably, of ploughs, drill ploughs, horse breaks, &c. This induced him to be very voluminous in their description, and very hyperbolical in their praise. Such passages I think are on the one hand almost unintelligible for want of plates; and on the other hand very suspicious, from being in the sale a source of profit: they are further much inferior to similar inventions that have arisen since;

PREFACE.

since; so that it would be swelling these pages prodigiously to insert all his accounts of this sort, without adding any thing to their value.

Another part of his works which is left out of this edition, is his recommendation of steeps, liquors, receipts, nostrums, &c. some of which he sold, and consequently wrote of them with a view to increase the sale; others are useless, and all too unimportant to obtain a place among the better exertions of his understanding and his practice.

I have likewise been cautious of admitting the drilling and horse-hoeing parts of his books; it is evident that his own practice was not the foundation of his ideas, and his praise of the new mode was too connected with his advantage in selling implements; besides which, later writers have so much exceeded him that the reader suffers no loss.

But in stating and explaining the reasons of good common management, the case is

far

far different from all thefe: we then plainly perceive that he knew what he wrote of; he fpeaks clear, found, good fenfe, and fo much to the purpofe that I will venture to affert no writer has in this path exceeded him: but the public have paid fo little attention to his works from the quantity of rubbifh they contain, that *Ellis's* real merit is little known. It is not at prefent recollected, that all the fpirited practices of excellent common hufbandry, which have of late years made fo much noife, are clearly afcertained by him, their merit ftated, and their conduct explained. The beft turnip and clover hufbandry are in particular fet forth, as practically as they can be at this day: the whole conduct of manures, though not philofophically handled, yet are ftated with practical precifion; and the common management in them fully explained. A full knowledge of the ufe of foiling cattle with tares, clover, &c.—faving the drainings of the farm yard—forming compofts—the variations of foils which require correfponding variations of manure and tillage (an article of great importance, and fully treated by no other writer)—

and

and the whole management of sheep—are among many other instances of *Ellis*'s thorough knowledge in common husbandry.

Respecting the mode of registering his observations, it is nearly retained; for I do not conceive it fair to alter the turn of any passage, the only liberty I have taken is to draw a pen through what I think trivial or exceptionable, and to change the heads of his paragraphs, which all begin, *how* this happened, and *how* such a man did so and so; by leaving out the word *how*, and giving only the subject of the paper. As to the cases he states in the practice of his neighbours, I apprehend the method quite unexceptionable, and accordingly retain it with the names of the persons.

Having premised these circumstances, I shall no longer detain the reader, but take my leave with observing, that the writings of this plain country farmer should not be held cheap, because they abound with none of that shewy science which glitters in the works of some other authors on this subject; such works may meet with great applause
from

PREFACE.

from those who are satisfied with nothing but philosophical researches, and having been accustomed to this garb of most other subjects, must have their philosophical farming too; but the true practical man, whose income depends on well tilled fields, may in the simple page of this husbandman, meet with more valuable hints than in all the philosophical treatises ever written. Some of which on the subject of agriculture have lately appeared, that are so very learned that none but chymists and mathematicians can understand them; such works may be very fine things, but perhaps they may be too fine to do any good. *Ellis* has his merit, though he is neither a chymist nor a mathematician.

CONTENTS

OF THE

FIRST VOLUME.

BOOK I.
Of SOILS.

CHAP.
I. Of Clay — page 1
II. Of Gravels — 12
III. Of Chalk — 25

BOOK II.
Of MANURES.

I. Of the Sheep Fold — 28
II. Of Farm Yard Dung — 35
III. Of Urine — 40
IV. Of Composts — 43
V. Of Chalk — 49
VI. Of Marle — 55
VII. Of Lime — 57
VIII. Of Burnt Clay — 60
IX. Of Peat — 65

CONTENTS.

CHAP.
X. Of Coal-Ashes	p. 74
XI. Of Wood-Ashes	77
XII. Of Soot	81
XIII. Of Malt Dust	88
XIV. Of Salt	ib.
XV. Of Oil Cake	90
XVI. Of Horn Shavings	91
XVII. Of Woollen Rags	96
XVIII. Of Crag	99
XIX. Of Sea-Weed	100
XX. Of Hair	101
XXI. Of Pigeons-Dung	ib.
XXII. Of Rabbits-Dung	104
XXIII. Of Dogs-Dung	105
XXIV. Of Human Ordure	107
XXV. Of Lime	108
XXVI. Of Hoofs	113
XXVII. Of Hogs Hair	114
XXVIII. Of Buck Wheat	ib.
XXIX. Of Old Thatch	117
XXX. Of the Comparative Quantities	118
XXXI. Of Manuring Meadows	119
XXXII. Of the Variation of Manures	120
XXXIII. Of the Importance of Manuring	123

BOOK III.
Of TILLAGE.

I. Of Ploughing	129
II. Of Rolling	143

CONTENTS,

BOOK IV.

Of the CULTURE of WHEAT.

CHAP.
I. Of Ploughing for Wheat - - p. 156
II. Of Wheat on Clover Lays - 168
III. Of Wheat on Sainfoine - - 170
IV. Of Wheat on Natural Grafs - 172
V. Of the Sort of Wheat - - 174
VI. Of the Time of Sowing - 191
VII. Of the Quantity of Seed - - 198
VIII. Of the Growth of Wheat - 199
IX. Of Feeding Wheat - - 206
X. Of the Smut - - - 216
XI. Of Harvefting Wheat - - 225
XII. Of Thrafhing and Cleaning Wheat 250
XIII. Of the Product of Wheat - 271
XIV. Of the Sample of Wheat - 273
XV. Of Rye - - - 288

BOOK V.

Of the CULTURE of BARLEY.

I. Of the Soil for Barley - - 292
II. Of Tillage for Barley - - 301
III. Of Manuring for Barley - - 307
IV. Sowing Barley on Natural Grafs 312
V. Of the Sorts of Barley - - 313
VI. Of the Time of Sowing - 319
VII. Of the Quantity of Seed - - 322

CONTENTS.

CHAP.
VIII. Of the Change of Seed - p. 323
IX. The Growth of Barley - - 325
X. Of Thrashing and Cleaning Barley 327
XI. Of Harvesting Barley - - 335
XII. Product of Barley - - 337

BOOK VI.
Of the CULTURE of OATS.

I. Of the Soil for Oats - - 338
II. Of Ploughing for Oats - - 341
III. Of the Time of Sowing Oats - 342
IV. Of the Sort of Oats - - 350
V. Of the Quantity of Seed - - ib.
VI. Product of Oats - - 352

BOOK VII.
Of the CULTURE of PEASE.

I. Ploughing for Pease - - 354
II. Of the Sort of Pease - - 360
III. Of the Time of Sowing - 370
IV. Of the Quantity of Seed - 375
V. Of Harvesting Pease - - 376

BOOK VIII.
Of the CULTURE of BEANS.

I. Of Ploughing for Beans - - 379
II. Of the Soil for Beans - - 385

CONTENTS.

CHAP.
III. Of the Sorts of Beans	p. 387
IV. Of Setting Beans	388
V. Of Hoeing Beans	390
VI. Of Topping Beans	391
VII. Of Rolling Bean-Ground	392
VIII. Of the Fly, &c.	ib.
IX. Of the Product of Beans	394
X. Of Harvesting Beans	395

BOOK IX.

Of the CULTURE of TARES.

I. The Sort of Tare	p. 397
II. Tares on a Sainfoine Lay	406
III. Of the Time of Sowing	408
IV. Of the Quantity of Seed	411
V. Of the Application of the Crop	ib.

BOOK X.

Of the CULTURE OF BUCKWHEAT.

I. The Advantages of sowing French Wheat	p. 429
II. The Culture of French Wheat	431

BOOK XI.

Of the CULTURE of TURNIPS.

I. Of the Soil for Turnips	p. 433
II. Of Ploughing for Turnips	435

CHAP.

III. Of the Sort of Turnip - - p. 448
IV. Of the Fly - - - 452
V. Of the Hanbury - - - 453
VI. Of the Time of Sowing - - ib.
VII. Of Hoeing - - - ib.
VIII. Of the Application of the Crop 455
IX. Of Saving the Seed - - 459

BOOK XII.

Of the CULTURE of RAPE, 461

BOOK XIII.

Of several CROPS not commonly cultivated, p. 482

BOOK XIV.

Of the CULTURE of CLOVER.

I. Of the Time of Sowing - p. 484
II. Of the Crops with which it is sown 486
III. Of the Growth of Clover - 503
IV. Of Making Clover Hay - ib.
V. Of the Application of the Crop - 505
VI. Of the Seed - - - 514

ELLIS'S

ELLIS'S HUSBANDRY.

BOOK I.
OF SOILS.

CHAPTER I.
OF CLAY.

RED CLAY agrees with almost all sorts of dressings, and is not of that ravenous nature as gravels and sands, &c. are; and to say the truth, less dressings and more ploughings, best suits this sort of clay, which I have often seen, returns its dung to view that had been ploughed in a year or two before; but above all dressings, none agrees so well with it as dry ones, such as lime, chalk, soot and ashes.

White clay, or tobacco-pipe clay, is not so tough, cold nor moist as the red clay, but has looser, warmer and drier parts, which render it a beneficial earth: this land will be brought into a condition for sowing of grain with less ploughings, and when fully encouraged by good tilths and dressings, will yield vast crops of grain, as may be annually seen in some parts of *Iving-boe* parish; this by the more able and better sort of farmers is double dressed by fold and cart dung, and frequently returns 40 bushels of wheat on an acre; and in this do barley and pease mightily thrive, as well as wheat, especially in a wet summer, but beans do not so well answer.

This flat white earth has, besides its many good qualities, one very bad one belonging to it, that often ruins part of its crops, while in their youth, and that is by frosts, winds, and rains. The frost sooner shatters and crumbles this earth than any other, that I know of, which in stitches or ridges, makes it fall from the roots, and leaves many of them bare; this evil is also much increased by the winds and rains, for in the winter and spring, the violent winds that succeed the frost, blow this sort of earth from the roots of the grain, and what that misses, the rains compleat, by washing the earth away from many of the roots; therefore the fold is here the very best remedy, which by the tread of the sheep, makes this ground turn before

before the plough in a clotty substance, and then it is in that order the farmer likes best.

For as this white land is easily brought into a tilth, and so becomes finely powdered, it produces vast quantities of poppy, which is the chief weed that hurts this ground, and then the farmer is like to suffer; but the more to prevent that, and also the damage that often happens by the frosts, winds and rains, they commonly sow this ground before, as well as at *Michaelmas*, with wheat, and run the fold over it afterwards, in order to settle and fix the earth about and upon its roots, that the weather may not have power to spoil the crop.

Again, after this they use the roller, by drawing it across and along the stitches, to fasten and inclose the roots of the wheat, which makes a second and double security against the wind, rain, sun, and frost, and is much better than top-dressing the grain after it is in the ground with dung, by reason the dung will by its heat keep the ground in a loose condition, and so hinder the roots from taking fast hold of the earth; this makes the fold on this earth preferable to dung, although it is ploughed in, and also to light hand-dressings that are sown early in the spring on this ground, as soot and the other sorts. In this white clay, clover will grow, though not so well as in the red, and yellow, and black clays; and therefore trefoile and fainfoine are generally here fown before it; thorns and other hedge-wood will grow well in this foil, and fo will elms.

As this land is of a warm and fomewhat dry nature, they fow their wheat, and their *Lent* grain early, that they may get cover before the winter frofts for the wheat, and fhelter againft the fun's heat in the fummer; and more effectually to come by an early and full clover, they fow more feed in this ground than ordinary, to fhade the roots; and even about *Chriftmas* I have known them to fow peas, fo that great dreffings, and thick early fowings in this fort of white ground, are perfectly neceffary; but particular regard is generally had by all occupiers of this white ground, to endeavour the fowing it in wettifh weather, that this ground may turn up clotty before the plough, whereby the feed is the more faftened in it, the poppy prevented, and the crop better fecured.

Yellow clay requires ploughing in much the fame manner as the red clays, and as it is more loamy than the red, it is certainly the richer foil, and in it both trees, grains, and grafs will grow and thrive much fafter; fo that as this is the better fort, I need fay little more, than that the fame culture and management, that is requifite for the one, will do for the other.

Black

Black or blue clays, are the very best of all clays; it is this and the black clayey loams, that impower the fertile vale of *Aylesbury* to vie, I am of opinion, with the richest lands in *England*; and it is from hence, that the vale of *Esham*, *Rumney* Marsh, and other low grounds, furnish this nation, as from a magazine, with the greatest quantities and best sorts of wheat, beans, beef and mutton, &c. This black earth produces the best and finest red *Lammas* wheat, and in such large bodied corns, as intitles their sellers to 6 *d.* a bushel more than the chiltern men can generally get for theirs, although of the same sort, because theirs will outweigh ours by a considerable deal.

However, I have now the satisfaction to say, I hope in a few years our chiltern country, by the help of sowed grasses and turnips, will be able to get better crops of wheat and other grain than heretofore, and little inferior, if not as good as the vale. This black earth is composed of a black, blueish clay, with some black mould, and as the latter is more or less in it, it has so much a proportion of the loamy nature; it lies there of a considerable depth, free from stone, and clear from the great deep hogweed, cat's tail, and other trumpery, that the chiltern men in many places at this day are, through ignorance, troubled with, to the destruction of many of their crops, as I shall further make appear hereafter.

There is no earth in the kingdom will carry greater burthens of corn and straw on it than this will, and with as little dressing as any; for indeed the main kind of this black vale earth, is all a sort of marl, and though sometimes of a stiffish nature, at others it is as loose, by reason of the frosts and winds great power in bringing it into a crumbling, short, loose condition, so that here they commonly plough their very wheat in as well as their beans, for the sake of making it stand fast in their great open fields, their irons wear little, their weeds only thistles, docks, and hale-weed, their dressings cheap and on the spot, and their rents and servants wages proportionable; but then these conveniences are not without their inconveniences, for they sometimes, I may say commonly once in about three years, lose whole flocks of sheep by the rot; their lands often flooded, their horses heels frequently cracked by the dirt, their firing scarce, as being obliged to burn the stubble of wheat, barley and beans, instead of wood; and in short, they had need have something extraordinary to make them amends for living amongst mud, dirt and water, where not only their wheel carriage is confined good part of the winter, but also themselves from visiting even their near
neigh-

neighbours; besides the almost constant and greatest misfortune of all, of living in an unwholsome, agueish air. It is true, they abound in plenty of corn and the best of grass; but then they are strangers to that grand conveniency of enjoying fields of turnips, and sowing foreign grasses, by which, if rightly improved, we may make our chiltern as productive as their best grounds in the vale; and indeed it is this that is my ambition and aim, to study and find out by experience, that art, which will so help nature as to make poor land rich, by impregnating it with those fertile salts, and nitrous, sulphurous qualities, as to render it in effect equal with this black, marly clay, which I am persuaded may be done in numberless places, in this our hilly country: not that I pretend to say a chiltern man can farm as cheap as a vale man; that is morally impossible, from the hard texture of our earth, and the stones that are therein, which ever will be the cause of a greater charge of smith's and wheeler's bills, &c. and a larger expence of hand-dressings to supply the hungry, lean nature of our high and dry grounds: but then we have our advantageous returns, of turnips, sowed grasses, firewood, and other sorts, besides the enjoyment of the third year (that with them lies fallow) under such feeding which they cannot have, because on black, clayey mould that is deep and wet, neither the turnip, nor artificial grafs will anfwer, by reafon of the cattles ftolching, nor will the turnip any more than the cherry or beech thrive in their wet, fpewy, clay ground.

The Culture of Clays.

The red clay being the moft obdurate fort, and the moft difficult to reduce to a fine tilth of all the reft; the art and labour of the moft accurate ploughman is fometimes foiled in his endeavours to accomplifh it, by reafon this earth is made worfe to anfwer that end, by the extremes of wet and drought. By wets it becomes more faft and tenacious than at other times, as lime is by water converted into a more vifcous, binding nature; fo that when the clay is in this condition, the coulter cuts through it without breaking the clots, which caufes fo little impreffion and alteration in its body, that oftentimes three ploughings in fuch fort, are to no greater purpofe, than one in chalks, gravels or fands; this has been fo often manifeftly proved, that notwithftanding all the fallow feafon has been employed in ploughings and crofs ploughings, and as frequently harrowed; yet in a wet fummer I have known it out of the farmer's power to get this ground into a fine tilth, which for all their dreffings of dungs and foot, has spoiled their crops of wheat, by the black-bennet, horse-gold, and other weeds growing amongst and crippling the corn, which is the sole occasion that chalks and sands are made use of, to alter its stiff and sour nature; and this it will effectually do, if the chalk is good, timely drawed, enough put on, and rightly ploughed in.

By droughts this clay is hardened, and when ploughed in this season, arises in great clots, that will not yield to the coulter's cut, nor the share's break, but will rather suffer driving up together in clotty heaps, than breaking into a small body; and therefore it is that the red clay when under such an untoward texture of body, is suffered to lie exposed to the weather, till a good shower of rain falls that will slacken, soften and meliorate its hard parts, and make it fit for reduction by the harrow-tynes, and then no time should be lost in taking this advantage of the clay's alteration, by harrowing it thoroughly both ways: for, as I have generally observed, this evil of a sour tilth is partly occasioned by want of timely ploughings and harrowings according to our maxim; clays cannot be ploughed too often, nor gravels, chalks and sands too seldom. Therefore if a wheat stubble ground is to be got in tilth for peas' against the next spring, then ought the stitches to be
ploughed

ploughed up into bouts at *Albollantide*, and left all the winter; then before you sow the peas, harrow it, and sow them in four thoroughs in the same stitch in the method it first lay, then after the pea or bean is sown, harrow it. Some will plough the stitches into broad lands at *Albollantide*, and let them lie all the winter, and harrow the latter end of *March*, when they plough acrofs the broad land and sow in four thorough stitches. The other sort of culture requisite for red clay may be the very same as is set down in the culture of loam.

Yellow clays call for much the same husbandry.

Black clays are different, because this earth being of a shorter nature than either red or yellow, will yield much sooner to the operation of the plough, therefore fewer ploughings will here do more service, because this earth must not be too fine when sown, for the foregoing reasons: formerly they sowed this ground always in broad lands, but now they sometimes (but seldom) sow it in stitches for wheat, but nothing else. The broad lands hold water more than a stitch, and the sun cannot dry it as it will a stitch.

BLACK CLAYS are a medium soil between the white and red clays, as to the ploughing part, and therefore too seldom or too often plough-

ploughings are equally a fault here. This earth lying moftly in vales, is commonly ploughed with the foot plough only one way, in broad lands, by reafon they lay it in half acre pieces, and that as high as they can, to avoid the wet's pernicious confequence; fo that here they directly draw the plough up and down by dividing the half acre, and turning the land two contrary ways, by which means a henting or large thorough is left in the middle.

In the large and fertile vale of *Aylefbury*, as well as in moft other low grounds, their earth is generally of a black and bluifh clay, or a marly dark coloured loam, clear of ftones, which in winter is very apt to ftolch by the tread of cattle, and by the wafh of frequent rains that eafily converts it into a muddy confiftence, and in frofty weather from a ftiff clot, it is prefently reduced into a fhort crumbling loofe body, which readily lets out the fpiry blade of new fown corn; alfo in this their low fituation, they are very apt to fuffer by floods, which fometimes are fo long retained on their ftiff foil, that the vale appears almoft like a little fea, ruining their grain, and rotting their fheep: on the contrary, in the fummer time, their open fields are foon dried, and their clayey furfaces become fo hardened, as to cleave afunder much more than any other fort of land, whereby their

barley

barley in particular is often stunted in its growth, and dwindles into a short, lean, thin crop, on their high, exposed, ridged lands; which at first was invented for the security of their growing corn against overflowing waters.

In the vale of *Aylesbury* and many others, their ground for the most part lies in ridge half acre, and whole acre lands, which are never ploughed across, but kept up to their stinted breadth, length, and height, according as they lie wetter or drier: and, as their soil is generally a bluish clay, or stiff black loam, nothing better answers their purpose than wheat and horse-beans. But as all ground naturally affects change, of late many have sown a tilth crop of barley, where wheat and beans have been before, and found it succeed to their wish. Pease they seldom sow, because this rich earth is very likely to cause them to run into straw, and too little into corn.

CHAP. II.
OF GRAVELS.

GRAVELS. This earth has variety of natures according as they severally abound with diversities of earths, or stones. There are the sharp or stony gravels, loamy gravels, clay gravels,

Chap. II. OF GRAVELS.

gravels, sandy gravels. On all which I shall distinctly make my remarks, as they have occurred to my knowledge; having all but the sandy sort more or less in my own fields.

Sharp Gravels.

This earth is composed of small, sharp stones, mixed with some mould, which is better or worse, as the same is in a less or greater proportion: if more stone than mould, then it is so much the worse, because the soil is so much the looser; if almost all stones, as some is, it is so much the worse still, as some is near the top of *Dunstable Downs*, and in many other places in this county, not worth above 1 *s*. or 2 *s*. 6 *d*. an acre; because the water runs so swiftly through them, that it washes the dungs, or other dressings, away below their roots; so that the fibres of corn or trees soon become hungry, pine, and oftentimes perish for want of the dung's nourishing subsistence; that in clays and loams have firm and holding bottoms, and will lie two or three years to visibility. There is also another sort of hungry gravel, that is made up chiefly of the round, blue pebbles; this of all others is reckoned the worst, and of the most cormorantine nature; this is such a devourer of dung and other dressings, that in many places they utterly refuse to plough or sow it.

it, but let it run with what poor grafs nature in a small degree throws up.

Of this last sort there are great quantities about two miles from me, which might be vastly improved, did not the tenant's obstinacy prove a barrier to his interest. I was lately concerned at an appraisement of a crop of several sorts of corn that grew on such sort of ground, as these two are. Some of the wheat at harvest was dear in my judgment of 20*s.* an acre, notwithstanding it was dressed, which helped to break the tenant, and obliged the landlord, who was but in indifferent circumstances himself, to take it into his own hands: I told him, to improve it to a good account, was to dress it with *French* wheat, but I cannot understand my advice had any more effect, than if I had told a passant traveller of a mine in *India*.

One farmer said to another, *Fear not, if the stones are big enough, you will get a good crop.* For, when they are small, round, blue pebbles, it is a sure sign of a very hungry poor soil.

Loamy Gravels.

Are those of a better sort than the former, for under these terms as much or a greater quantity of mould is supposed to be, than small stones; this I may justly term a fertile soil if rightly managed, because here is some hold for
the

Chap. II. OF GRAVELS.

the roots of trees, corn, or grafs to receive nourifhment from: and as all gravels are of a kerning, or corn-yielding nature, when fupplied with heaven's propitious influences, and man's due affiftance; I have known this ground to return above eight quarters of barley on an acre, and other grain proportionably.

Moft gravels are of a hot, dry, loofe, hungry, and binding nature, which oppofite terms, though feemingly inconfiftent, yet carry experimental truths with them. Loofe, dry and hot, I fay, for that this compound ground of ftones and mould cannot by their make and texture fo unite, as to be a clofe body, by reafon of the various fhapes and bignefs of the many ftones with which the mould is conjoined, as a fmall one amongft two great ones, or angled ones among globular; and fo the reft are of fo many different forms, that muft caufe a hollownefs and disjunction of parts with the earth, notwithftanding the moulds enveloping many of their irregular fides; from hence it is, that the rains and fnows fo eafily make their paffage through the cavities of this ground, to the great prejudice of the trees, corn, or grafs that grow therein. Hungry; becaufe the dungs and dreffings are fo readily carried off, and wafhed away paft recovery; that without more than common fupply of mould and dung, this will (I am of opinion)

prey

prey on its own mould substance, and so grow leaner and more barren, by sowing its most vital part the mould. This leads me to take notice of the notion of some, that think these small stones breed and are nourished by the sun, air, rain and mould; and indeed it seemingly carries a probability with it, for it is surprising to find hardly any diminution in a small field, where vast quantities have been carried off to mend the highways: if so, then by consequence the mould administers great part of their sustenance, which perhaps may be one reason why these gravels are so voracious.

I know a great gentleman that now wants several of the great stones that we call growing stones, composed of vast numbers of small pebbles that lie in little cells or holes; his use for them is to put them down in his grounds for lasting boundary marks to the several parishes, that triennially make a progression, and cuts deep and large marks in his great trees for their future knowledge, how far their parish limits extend, to the damage of the said trees, which he hopes to prevent by placing these sort of growing or ever durable stones at the proper places. And possibly, nay likely, they may deserve those names, though imperceptible to our eyes or memory.

These gravels are binding; this needs no other proof than the many misfortunes accru-

Chap. II. OF GRAVELS.

accruing by its hard and crusty surface, so made by the weights and bashings of the heavy rains; and more or less so, as they sooner or later succeed the ploughings and sowings; for if they come before this sort of earth is settled by time, it will harden it the more, and sometimes cause it to run as it were into a pancake spread, particularly on descents, and bind the wheat, peas or barley in so fast on all sides, that they are often ruined, as not being able to make their growing progress, for want of room in the earth, for the swell and multiplicity of their several stalks; and this I take chiefly to be occasioned by the rains washing the mould on the stones, which naturally joins and cements their hard bodies to the moulds soft, plastic nature, and thereby becomes one close and obdurate ground.

A Case.

A field near me had a gravelly, pebbly bottom, covered by a black, light mould of about four or six inches deep; this was a lay of natural grass about twelve years ago, since which it has almost constantly been under the plough: this sort of ground is naturally called a dying ground from its great lightness, because its earth is very hollow and its bottom retains no wet; so that though it was well ploughed, well dressed and a flourishing crop of wheat with large ears, yet it moſtly died, and the weeds, as the May-ſeed, horſe-gold, poppy, and wild oat got up to that degree this wet ſummer, 1732, that it was hardly worth reaping; and indeed it is the nature of moſt grounds, loams and clays eſpecially, to produce the wild oat if ſowed too conſtantly and too long with corn.

The remedy of this is, to give it conſtantly its due courſe of fallowings, whereby it may enjoy a thorough ſweetneſs; and let the dreſſing be every wheat year or barley ſeaſon, a manure accordingly; in the wheat year the fold is certainly the greateſt friend to this ſort of ground of all others, becauſe it will tread this light earth, and bring it under a cloſer texture of body, than otherways by its own nature it would be; and therefore far better for the crop of wheat that is to grow in it, and more diſcouraging to all manner of weeds and worms that will be ſure to infect this light ground, eſpecially in wet ſummers, if care is not made uſe of to prevent their ill conſequences; beſides, the fold by the preſſure of the ſheep's feet and bodies enables this hollow land to ſuſtain and hold the roots of the wheat faſt, that the winds and rain cannot ſo eaſily force it down, which they will certainly do when that is ploughed often and ſowed in a fine duſty tilth, and this more ſurely

if

Chap. II. OF GRAVELS.

if manured with dungs of horses, cows and swine, that increase the ground's lightness and keep it hollow afterwards. It is therefore that this loose earth and dry bottom should be ploughed and sowed in a wettish time, for that contributes vastly to its binding, and causes it to remain so during the next winter and summer that the wheat is to grow in it. As to barley and other seasons at the spring of the year, the case is somewhat altered, by reason the wets that generally fall then, help to the fastning of this ground; that may therefore be ploughed and sowed earlier than clays in order to enjoy the same; besides, the *Lent* grain has only about half the time of the wheat to be in this earth, which gives not that opportunity to weeds for their predominancy as the long wheat season does: but such ground as this is best laid down with artificial grass, in order to obtain a natural sward, which this earth will naturally run into, and much quicker if encouraged by sowing the seeds of fine upland, meadow hay amongst the clover, trefoil or ray grass. The occupier of this field, notwithstanding he has often ploughed and sowed it, was ignorant of the true nature of this soil that lay contiguous to three other fields belonging to the same person, as believing it to be of much the same nature as the rest, and therefore gave it the same usage; but herein he remained miftaken, till a perfon of better judgment than himfelf convinced him of his error.

All which plainly fhews the excellency of this knowledge in the nature of earths; and as it is in the animal oeconomy, that there is no right application of a remedy without the difeafe is firft known; fo it is here that every farmer ought to have made it his primary ftudy to inform himfelf of the feveral forts of ground that often belong to his farm, and that befides his own judgment to confult his neighbours, who as natives on the place may be able to let him know more than the dictates of his own reafon, that formerly were more remote from the fame.

Clay Gravels,

Or clay, loamy gravels, happen to be part of the ground of feveral farms in this country: their nature varies but little from the gravelly loams, but wherever the red clay is part of the foil, there will be occafion for more ploughings, and warmer dreffings than in any other fort of gravel.

Sandy Gravels,

Are of fo loofe a nature, that they are fooner brought in order than any of the reft, and will bear very good crops of corn, if duly affifted with manure; and that much forwarder than the

Chap. II. OF GRAVELS.

the other gravels, whereby peas and turnips may be had in one and the same year.

The Culture of Gravels.

Sharp, or loamy Gravels require much the same sort of ploughing and harrowing, as also the same sort of dressing; these as well as the sandy sort are in the number of the light, sweet soils, and are all of them so different from clay, that too much ploughing here will wear out the ground, as too little in that soil will not bring it into a bearing condition: a good season of fair weather is more than ordinary requisite in these soils to prevent its binding; and also shallow ploughings when grain is sowed on broad lands on the same, otherwise it is in great danger of being hindered getting through.

The clay gravels are still more binding, and therefore must have more ploughings.

Gravels, of all other sands, stand most in need of being brought into a condition, as will strictly answer this title; to find out which, in a true beneficial manner, many have been the attempts, and various efforts of farmers; which as it is a matter of great consequence, I shall here mention several particulars. First then, I knew one, whose farm of about 60*l.* a year was chiefly a sharp and loamy gravel, but not of the blue pebbly sort; this man carried

from

from his farm juſt by me ſeveral waggon loads of peas, thetches, chaff and other grain to *London*, in order to load back again with coney-clippings, horn-ſhavings, ſoot, &c. This dreſſing did not the firſt year do quite ſo much good as afterwards. It happened by the landlord's diſguſt, that the man after laying out great ſums on this ſort of dreſſings, was forced off his farm, and ſucceeded by another good huſbandman, who directly enjoyed the former's expence of dreſſing; for this ſort of manure is not eaſily devoured by the gravels, nor waſhed away, as being of a tough, ſpungy nature; ſo that it will lie and hollow the ground, retain the wets, and ſo keep the ground moiſt and warm for ſeveral years. To this was joined another expence, and he was the firſt man in theſe parts that chalked gravels to the wonder of the other farmers about him. However, this anſwered its full end in all reſpects, for it abſolutely hindered the gravels from being cloſed and bound by the rains or ſnows, added a more loamy part to the ſtony part, made it plough much better, and kept it in a pure, ſweet condition, that has for ſeveral years bore extraordinary good crops of all ſorts of grain that grow thereon; but the uſe of theſe ſort of light dreſſings from *London* are much more laid aſide, and leſs regarded than formerly, by reaſon of the great numbers of ſheep that

are

Chap. II. OF GRAVELS.

are kept and folded on thefe gravels and other grounds, that are found far cheaper, and I am certain, are quicker and better dreffings.

Some will in lieu hereof lay on and plough in their long horfe litter, allowing it to anfwer beft in gravels, as being of the horn-fhavings nature, tough and fpungy. Others will lay on mud, or highway ftuff, which indeed has vaftly enriched this fort of ground, efpecially if it is a true mud, free from fand. For fomething muft be done to thefe gravels by way of dreffing, or elfe nothing but poverty will fucceed in this hungry foil; and to fay the truth, if any ground ftands in need of double dreffing, this does; and then there is none will pay better, by returning the beft of crops from its kerning quality.

But the horfe litter will anfwer very well another way, that is, by laying it on the top of the ftitches on broad lands, as foon as the wheat is fown (for then it will grow thorough it.) This is an excellent way to plough fome in; and afterwards lay fome all over on the top; for as gravels in general are a light, loofe ground, this cover will preferve it from fhoaling in the frofty feafons; and before next harveft it is a rarity if it is not hauled and pulled into the ground by the worm; or elfe devoured by the voracious nature of the gravel, which make potatoes, and all of the haulm tribe, to be fo good dreffing for

this

this soil.—It is strange, at first sight, to see great crops of wheat and other corn grow (seemingly) amongst vast quantities of stones, that in this country are common to be seen, where hardly any mould can be discerned.

But this is accounted for by the owner very plainly, when he tells the querist, that he would not give one load of stones for several loads of dung, because these stones have several advantageous properties in them: first, as they are of a cold, moist nature, they preserve the roots of grain from being dried and scorched by the great heats: secondly, they help to keep in the *Sal terræ*, vapour of breath of the earth, which by their cover is obliged to perspire more slowly; nor is it so readily exhausted by the sun's attraction, and therefore administers its fertile quality more regularly and more abundantly to the vegetables that grow amongst them, in that little mould there is, which the small fibres of the roots will be sure to search for and find out, and join, although but in a very small quantity, and that lodged amongst the several crannies and cavities of the stones.

Ragstone Gravel.

To define the nature of this sort of earth is what my reader, I suppose, expects; and therefore I have to tell him, that this is a thin, short, chalky

chalky surface, which commonly lies on a hurlock or ragstone; that is to say, a whitish, hard substance, between a chalk and a stone, which is of a most hungry nature, readily receiving and consuming the assistance of any manure, by letting the juices, or wash, easily run into and through its many joints, and hollowish body. Of this soil, there are great numbers of acres in *Bedfordshire, Buckinghamshire, Hampshire, Wiltshire, Dorsetshire,* and in other parts of *England.* Near me, there is a poor sort of this *Sugar-Plum Land,* as they call it, lying in the open common fields under *Dunstable Downs,* so poor, that the tenants are obliged to let great quantities of it lie two years together fallow or idle; in which space of time, it throws up a little natural grass, that serves as picking to their folding sheep.

CHAP. III.
OF CHALK.

THIS soil, the dry, lean sort especially, being of a short, crumbling nature, is easily got into a tilth, by reason on this ground weeds grow the least of all others whatsoever; and therefore two ploughings in this sort, will do as much service as four will in red clays, which

makes

makes the country-man say a fallow and a half is enough for a chalk; for here is not a conveniency for bouting and four thoroughs, &c. as in other lands, which obliges the ploughman to turn it each time of ploughing, if he can, the reverse cross way of the last operation, and generally into broad lands, which formerly was altogether the method, but latterly some sow their wheat in stitches. And let what grain soever be sowed in these chalks, it ought to be sown in wettish weather, because it is then made something clotty and rough, which best hinders the growth of their only and most pernicious weed the poppy; and also by its binding quality caused by the wets, it is better fastned, and will stand the frosts and colds much better: whereas a fine tilth in this soil is altogether to be rejected; for as the chalks are naturally light and loose in themselves, they are made so much the more so by being fine, and then the wheat and other grain will fall before the high winds, and sometimes be almost spoiled.

The marly clay sort indeed will give more room for the plough, because it will admit of a greater depth; and the loamy chalks more room than the last. However, be it ploughed in broad lands or stitches, rolling is very necessary, especially in the last, where it is used by being drawn long and cross ways, in order to fasten

and

and keep the ground firm and close to the roots of the grain; and also the better to preserve them from the violent heats and droughts in summer, which broad lands particularly are more capable of doing than stitches, as they lie flatter, lower and more solid.

The worst of these chalks that I call of the hurlucky or stony nature, will bear, when thorough dressed, good wheat, rye, barley, peas, thetches and lentils, but where this dressing is wanting, they commonly sow for *Lent* grain. Lentils and thetches, these both will grow and flourish in the poorest chalks, as may annually be seen under *Dunstable Downs*, where these two are often sown amongst oats, as bullimon, as being surer in their returns by far than peas,

BOOK II.
OF MANURES.

CHAPTER I.
OF THE SHEEP FOLD.

SECTION I.
Excellence of this manuring.

THIS dung is certainly the more rich, as it is made by a beast that can and generally does subsist without the help of water all the year; for which reason, their dung is more than ordinary impregnated with those vegetable salts that are contained in the grass, hay, straw, and corn that they alternately feed on: it is this saline excrement that is an enemy to all worms, slugs, grubs, flies, and caterpillars, as we find to our great profit, by folding before and after the sowing of turnip, wheat, barley, and other seeds; and when fed on hay or straw in yards in the winter, these creatures are of prodigious service in converting stover to one of the best

of

Chap. I. *Of the* SHEEP FOLD.

of dungs; which have partly occafioned many of our farmers this great ftraw year, to buy in wethers to eat it and make dung inftead of cows, that fold as dear as sheep were cheap.

The ftale likewife of this beaft is of courfe hotter and falter than fome others, by fo much as it is lefs ufed to fupply its drought by drinking; and therefore it is of a more fertile quality to the ground, which caufes the diligent shepherd every morning to drive the sheep brifkly about the fold, that they may be provoked to dung and ftale, before they are let out.

The dung and ftale of sheep are moft efficacious on all manner of loams, but more on gravels, chalks and fands; becaufe thefe grounds being of a light, hollow nature (efpecially the two laft) are by the help of the fold, brought under a clofer union in their parts, than they otherwife would be, by any other method that is now in ufe, whereby this dreffing is made to mix with, and ftick to this loofe, fhort earth in fuch a tenacious manner, as enables it much better to yield its foil to the corn.

As to the dreffing of land by sheep; while they are penning and dunging it, they do at the fame time rather prevent the breed of worms, than increafe it; for no reptile can agree with the urine of any beaft: and this I take to be one of the ftrongeft forts, as being fomewhat of the nature of their ftrong-fcented wool.

Secondly, As sheep are their own porters, they carry their dung and urine to distant fields, free of any other charge than a shepherd and his dog, where they have their grazing on commons and in fallow-grounds.

Thirdly, The dressing of sheep may be made use of as an alternate one, and thereby gives the ground a natural and refreshing assistance; especially, when it succeeds the manure of cart-dung.

Fourthly, The penning of sheep gives many farmers a most valuable opportunity to get the best of crops of wheat, barley, and other grains, and turnips, and grasses, by the dung and urine of them; for it is a present practice both in *Vale* and *Chilterne* grounds, to fold wheat and barley, after the seed is sown; especially on that wheat-feed which is sown on a lay of clover, or natural grass, when only one ploughing (as the usual way is) is given it, and the seed harrowed in; then it is that these most serviceable creatures, by penning them on it, not only dress and enrich the ground, but tread in the seed, and so fasten it, that neither winds nor rains can blow nor beat down its stalks, nor blow nor wash away the mould from off the roots: whereas, without such their treading in the seed, as it lies in this loose shallow situation, the wheat would be apt to grow up and stand so weak,

and

Chap. I. *Of the* SHEEP FOLD.

and be thereby so much under the power of great winds and rains, that it is rarely known to escape falling down in its green ear; and then it generally returns the farmer not above half a full crop of this golden pay-rent grain: but whether such penning of sheep be applied to wheat-seed sown on lays of grass, or on wheat-seed just sown on tilth broad lands, it has the same effect.

So likewise, where this excellent piece of husbandry is practised on barley-crops, it answers the same profitable ends. Therefore, it is now a common practice, both in *Vale* and *Chilterne* countries, where a farmer can conveniently do it, for him to pen or fold his sheep, on his new-sown barley-seed. In this last spring-season of 1745, I folded my sheep on part of a barley-field; and the other part of the same field I footed, by sowing over it about twenty bushels of *London* soot on each acre: and though a rainy summer attended the crop, I believe I may say, that so far as the fold was set, the barley was near as good again as where it was footed.

SECT. II.

Season of folding.

IF we dress our barley-crop with the fold, it must be done only in a dry time, for, if the fold was to be employed in wet weather, it would

would do more harm than good, by the sheep's treading in the ground so hard on the barley, as to prevent a great deal of it ever coming out: but if the fold is set on the barley, directly after it is sown, and the folding or penning is carried on in dry weather, it is the best of dressing, for chalky, sandy, gravelly, and dry loamy earths; because the stale and dung of the sheep will add such a fertility to the barley-crop, as to make it become a very good one, indeed, for I never knew this piece of husbandry fail its owners, when carried on in a right manner; and this may be done for some time, in one and the same field, even till the barley is three or four inches high; and, though the sheep may eat and trample some down, yet their dung and stale will so revive it as to force it on into a most quick growth.

To dress always with the fold, I absolutely deny to be right husbandry; and I do affirm for truth, that whoever folds one piece of ground every year, and many years together, for getting a full crop of grain every time on it, will find themselves mistaken in their hopes: nay, I will carry the matter further, and proceed to prove, that if a person folds such ground only once in three years, and continues this custom many years, he will also find himself in the wrong of it; which I thus make out.

The

Chap. I. *Of the* SHEEP FOLD.

The dung and ftale of fheep is known, to all that make ufe of them, to be a moft thin dreffing of the ground; for although the dung and ftale of fheep adminifters a nutriment to the earth, yet it does not do it in fuch a plentiful degree, as to laft above one year to a good purpofe; and therefore it is we look on the wheat-crop to be improved by it.

But the next year's lent-crop has but little fhare in its fertility, becaufe we reckon its virtue lafts but one year; for, at beft, the dung and ftale of fheep, that eat nothing but grafs, is allowed to be but a cold dreffing to the ground; and that foot, which is the thinneft of manures, exceeds it in refpect of duration; for that this black dreffing is endowed with fuch fulphureous and nitrous qualities, as to affift the land it is laid on two years together. However, fure I am, that neither the fold, nor foot, nor lime, nor afhes, nor oil-cake powder, nor malt-duft, will anfwer a farmer's intereft, if he always dreffes his land with any one of them.

SECT. III.

Folding Meadows.

OTHERS, again, will fold on their meadow-ground for improving it to a great degree, as a late gentleman ufed to do in the parifh parish of *Studham* in *Hertfordshire*; who, as soon as he had done folding on his wheat-land, about *Allhollantide*, removed his fold to the meadow, and there began folding on the same; for here it does great service, because now neither the sun nor air can dry up the dung, nor exhaust its virtue; and it is now that sheep kill the present moss, and hinder the breed of more by their tread, by the warmth of their bodies, and by their dung and stale, which is readily received by the earth, as it is now in a soft condition; and which will be of more service, if some straw is laid every night on the ground where the sheep are to lie, and hay or straw given them in a rack. But, to do this in the most efficacious manner, some of the nicest farmers fold two nights together, on one and the same piece of ground; and then it will surely answer their end, and not hurt the sheep, provided they are not folded in too wet weather; and this is one reason why they lay some straw at the bottom of the fold, because it defends their bodies from the bare ground, which, in winter, is very apt to draw the heat out of the sheeps bodies, and lodge a prejudicial coldness in its room. When folding is hired, the common price is three-pence a score for sheep each night.

Chap. I. *Of* FARM YARD DUNG. 35

SECT. IV.

Folding in the Farm Yard.

A Covered fold is a very great enricher both of grafs and arable land, by folding on the fame all the winter with wether fheep, firft ftrewing every night a little ftraw over the ground to keep it from too violently drawing the bodies of the fheep, and will alfo occafion their making more dung when very great rains happened, or when fnows fall, and even then, he fhould not lofe the benefit of the fheeps dung and ftale, but in this cafe confine them in a ftraw-yard, and feed them there with pea, barley, or oat-ftraw, out of a rack (this was the practice of a farmer) fo that he made it his particular care to enjoy the benefit of this excellent dung all the winter in one place or other, and indeed I may fay, all the year *.

CHAP. II.

Of FARM YARD DUNG.

I Know only one farmer befides myfelf, even in the parifh where I live, but what expofes both their long and fhort dung to the wafh of rains,

---

* The Author has much other matter concerning folding, but only relative to the welfare and conduct of the flock, not the good of the land, it does not therefore come properly in here. EDITOR.

rains, by throwing it out of the stable into the farm-yard, and laying it all over the same in a thin condition; which consequently gives the descending waters an opportunity of washing away the corn, and best part of their dung; and this the more, as the weight of horses, cows, sheep, and hogs bodies compress and squeeze out the goodness or quintessence of such dung, and leaves behind little more than the husky, spungy straw-part of it for the farmer's use. Here is husbandry with a witness! I mean ill husbandry; for so it is in the highest degree, as I am going further to make appear, by asserting it to be too general a practice with some silly farmers, not only to suffer their dung to be laid in the farm-yard, and have its goodness washed out, but also to suffer this liquid goodness to run away into a common road or drain, so as never to be enjoyed by him; and this during the whole year.

A damage, I believe I may say, committed on himself in an infinite degree; for where a great farmer loses his dung in this manner, who can tell where the loss may end, as dung affects his land, not only in the next crop, but also in many afterwards? for as the produce of the first wheat-crop is, so is the second *Lent* crop better or worse; and so on, as these return more or less corn and straw, and them dungs. So their rich

swine-

Chap. II. *Of* FARM YARD DUNG.

swine-dungs are commonly little better husbanded, because these lie in their hog-yard likewise exposed to the wash of rains, to a great loss; and the more so, where many of them are kept, and their dung is made by the food of pease or beans, which contain more of the saline or sulphureous qualities than any other of their meat besides. According to what I have observed in several large farm-yards, I think I may say, those tenants have lost near a fourth part of the goodness of their dungs, before they have been laid on their fallow-grounds.

I remember the sight of a piece of ill husbandry I met with on this account, in *Aylesbury* vale, where, though the farmer rented a hundred and fifty pounds a year, yet his hogs I saw feeding in an open hog-yard up to their bellies in a standing black water, and this in the winter, during all the time they were fatting; so that they could never eat their meat without coming out of their stye into this water, that run into an adjacent great pond, and from thence, when it overflowed, into another large pond at a considerable distance, and there was partly lost. His fowl-dung was also husbanded in as bad a manner; for, though he kept the best part of a hundred cocks, hens, and chickens, yet he suffered as many as would, to roost on trees, and elsewhere, to the loss of their dung, which kept this farmer poor; but, as poor as he always was, he was careful to lay out some of his money in soot and pigeons-dung for manuring his ground; and, in doing this, he did well, but neglected his care at the fountain-head, for here he ought to have employed it first; that is, in saving his dungs in the best manner, as I am going to shew.

As soon as horses are taken into the house, the farmer ought to begin saving their dung in particular, after this manner: supposing him to have conveniencies accordingly, he may keep his short horse-dung by itself, and the long by itself, or both together under cover. My short horse-dung I lay in a place by itself under the roosting of cocks and hens, whose dung strengthens that of the horses, and the horses that, by preserving the virtue of the fowl-dung from being exhausted by the air; and to improve both, I oblige my maid-servant to empty the chamber-pot every morning over these dungs, which helps to rot them the sooner, and impregnate them with such a fertile quality, as renders them a most rich compost for manuring land.

And, when it is rotted enough, some sow it by the hand out of a seed-cott; others lay it in little heaps, and spread it before or after corn is sown, with a three-tine fork. As for the long dung, it must have a longer time to rot; but, when both short and long dung is mixt

and

Chap. II. *Of* FARM YARD DUNG.

and lain together, the short helps to rot the long dung much sooner. And if cow and swine-dung were thus incorporated together with horse-dung, and kept under cover, they would be *cent. per cent.* the better for it, in comparison of their lying all abroad exposed to the weather.

But, in case there are not conveniences for laying such dung under cover, then, as the dungs are made, they should be lain in one great heap or dunghill, which next to cover will preserve their good properties in a great measure from the power of rains and droughts; and, as the black water drains from it, it ought to be carefully preserved, by causing it to run into such a receptacle or reservoir, as will give the farmer an opportunity to carry it out in a tub or barrel, for throwing it over the dunghill, or to scatter it over plowed or grass-land, over wheat, beans, pease, or oats, in their infant growth; or to water cabbage-plants, or other culinary roots.

For, if stable or other dungs were laid thinly over the farm-yard, the rains would easily wash through them, and the sun dry them, and that much more than when such dungs are laid in a thick substance. But, before I quit this subject, I must observe, that I have seen a great farmer lay his stable-dung under a granary built high from the ground on purpose to be a shelter or cover for something. Here I should think it improper to lay dung, becaufe the fteam of dung is moft apt to breed a mould, that is pernicious to every thing it fettles, or gets to. When fowls-dung is kept by itfelf, as often as we have duft, offal chaff, or other trumpery, fanned out of the corn, we mix them with fuch fowls-dung, which, in time, will lie, heat, rot, and become an excellent manure, to be fown as I faid, out of the hand feed-cott, and harrowed in with your barley, or otherwife applied.

Where horfes are confined in no greater room than a large yard, one load of ftraw will nearly make ten loads of dung.

CHAP. III.

OF URINE.

THIS is allowed to be one of the beft dreffings for moft vegetables, if made ufe of at a right time, and in a right quantity. A farmer found it fo, when he ufed it fo late as in *March* on his wheat, which he fprinkled out of a garden watering-pot, over his wheat, as it grew in two-bout ftitches; and for this, and other purpofes, he faved it in a barrel which he ufed to carry into the field in a cart, and there drew it out into his tin watering-pot. But another farmer had a better contrivance; he kept his cham-

Chap. III.   OF URINE.   41

chamber-lie in a great oil jar, that is to be bought at the oil-shops in *London*, for about four shillings a-piece; when it was full, he put the urine into a barrel in a cart, and, in *January*, *February*, *March*, *April*, or *May*, would let it out into a wooden long trough, bored full of little holes, that lay across the tail of the cart, by which the urine would run gradually out over a great deal of wheat in a little time; and it was observed, that this farmer had the best wheat in the country on his chalky loams, near *Tring* in *Hertfordshire*.

Another trial was made by a groom who, by way of curiosity, had a mind to try the effect of horse-stale, and therefore sprinkled it over some wheat but once, that grew in two-bout stitches near *Nettleden*, which caused it to come on so rank that they were forced to cut it down several times, and at last it run so much into straw, that there was hardly any corn.

Another trial was made with chamberlie, by a gentleman who strewed it over the roots of a wall-fruit tree to make it prolific, but instead of that it killed it; however, part of the next tree's extreme roots having received some of it, it caused it to flourish in a furious manner, and bear more fruit than ever it did before. Another proof of the good effect of this was annually experienced by a gardener, who, having but a little spot of ground, enjoyed the greatest crops

of

of kitchen-greens in all the country about him, by the help of urine, which was his only dressing; and which he every year saved and sprinkled over all his land, and it caused his onions in particular, to come early, and grow into very large roots.

As it is endowed with a burning spirit, it calls for a careful management on all sorts of grass, corn, or trees roots, that it furiously assists or destroys, as it is discreetly or indiscreetly applied; its right use being not only in a small quantity, but at a proper season: in the first, it should be no more than sprinkled or poured on in small streams, out of a watering or other pot or dish: in the latter, it is to be done in *January*, *February*, or before *May* is over, that the drying heat of the weather may not add to the fiery part of the stale. I have thought human stale of that importance in promoting the growth of corn, grass, and trees, that I have and do allow my maid-servant threepence for every kilderkin she saves by emptying the chamber-pots into a cask, which when full, I put into a cart, and let it run out on my corn or grass, as it is drawing over them, by pulling out the cork at each end of the vessel. Likewise when I empty my necessary house, the man mixes straw with it in each barrow, and puts it on a heap to lie in the weather and rot, till it becomes fit to put on my land,

land, where when it is ploughed in, either at *Michaelmas* for wheat or rye, or in the spring for barley, it is of a very great and lasting service.

CHAP. IV.
OF COMPOSTS.

IF you have an opportunity before harvest begins, cut or dig, fork or shovel up grass-turf, that grows near hedges, or elsewhere; or where you have stocked up any brow of underwood: I say, mix such turf with lime, and it will burn up all grass, sedge, and small roots, and weeds, and thus reduce all tough, sour turf, and clotty earth, into a fineness and sweetness, against wheat season in *October* next, provided you can give the heap one or two turnings in that time. In the same manner all sullidge, and mud of highways, ponds, and ditches, should be served, either by mixing it with lime, or small chalk alone, or with lime and dung together; or with marle, maum, ore or sea-weed, fern, nettles, sea-sand, or other proper ingredients, to make a fertile compost.

This I have to say, in praise of some of the *Middlesex* farmers, about *Harrow*, *Stanmore*, and the adjacent parts, who make it their business to get a great deal of sullidge out of the bottoms of drains in roads, commons, and other places, which they here call a *mine*; for, with this sort and the emptyings of ponds and ditches, they mix *London* dung, *London* dirt, and small chalk, which some fetch, above five miles an end, from *Watford*; which after two or three turnings with some coal-ashes among it, becomes an excellent compost for grass or meadow ground, as well as ploughed land. The benefits of this husbandry are too many to be enumerated here, and, therefore, I shall only hint that, if ground is got into good heart, by being thus well dressed with a sweet, strong, and fine compost, the less and easier plowings will better answer the farmer's end, than more, and harder, without such improved compost-dressing; because it makes the earth plow and harrow much easier, and finer, than otherwise it would.

If the dryness of the weather will permit, empty your ponds and ditches in *June*, in order to give the most room for the reception of waters, and, at the same time, to enjoy a beneficial compost, which the mud, that is thrown out, will help to do, if rightly managed: for this purpose, let it lie to dry; when this is done, lay some small chalk close to it, and let it be mixed together into one square heap. Others will mix lime instead of chalk, which will burn up the seeds of weeds, take off the crude nature of the mud,

and,

Chap. IV. OF COMPOSTS.

and, much fooner than chalk, reduce its tough body to a fhort confiftence.

Or you may mix the fhovelings of dirt, or turf, or, if you are near enough the fea-fhore, ore-weed, or fand of any kind, together. All, or part of thefe, as conveniency allows, may be mixed, and, in two or three months time, turned and mixed again, and fo a third time, if occafion be, that all their feveral parts may be well incorporated. And, if fuch a heap can be got ready time enough to lay on juft before the laft ploughing for wheat, you may depend on it to be one of the beft of dreffings for almoft any foil, and will produce you a plentiful crop, free and clear of fmut, if other management is anfwerable to this excellent compoft.

Small chalk, or turf, or mould got from under hedges, or the fullage of bottoms in the roadway, or *London* afhes, or *London* dirt from off the layftalls, are what the *Middlefex* farmer chiefly makes ufe of for his meadow and ploughed lands; and, to give them their due in this refpect, I muft fay they are the beft of hufbandmen, becaufe many of them fetch their chalk, and *London* dirt and afhes, five or ten miles, and fo work and mix their heaps by feveral turnings, as to make them almoft fine enough to pafs through a fieve or fcreen, before they lay it on.

The fame character I muft alfo give the
Kentifh

Kentish and *Essex* farmers, who get all the turf, mould and mud, wherever they can, for mixing it with dung and chalk, or lime, in the greatest perfection. As to the quantity of one with another, there needs no computation from my pen, for every peasant is a sufficient judge of this, because there is hardly any danger of a mistake. I have a great hole in one of my fields, which contains, I believe, twenty loads of sandy mud, that it receives from the shoot of a hill once a year: this I have thrown out, and, when dry, I bring to it ten or twelve loads of chalk, and as much dung; or, if I put twenty of each of the two last, I do not know where the harm would be, as it is to be laid on a loamy gravel, for a crop of wheat or barley; for this soil, though short in its nature, yet, on the fall of great rains, when its in a fine tilth, will run together, like batter for a pancake; and when it thus happens presently after the seeds of wheat and barley are sown, it will bind them in so tight, that many of them can never shoot out; therefore chalk and other loose dressings are necessary here.

Mix mould with dung and lime; the latter will make the grazy or rooty mould run into small parts, and so foment them all, as to make them incorporate in a fine manner, and become fit to be laid on land, and ploughed in, in the month of

Chap. IV. OF COMPOSTS.

of *July*, for turnips or rye, or for wheat afterwards. Or if mud is so served, instead of mould, it will answer to a good purpose, and their proportion should be three parts mould or mud, and one part lime, or about ten loads of mould or mud, to two loads or ninety-six bushels of lime, to be laid length-ways, in a long, narrow heap, broad at bottom and narrow at top, somewhat like a hog's body, for the rain to wash the better off it.

This composition makes a most excellent manure for nourishing both corn and grasses; and therefore is very proper to be used on both ploughed and meadow-lands; and this advantageous practice is now become so much in esteem by the best of farmers, that, in order to come by this delicate dressing, they often dig up their mould, and carry it away for this purpose, from the very roots of their hedge-plants; and, to do this to the best advantage, some will first lay a quantity of mould or mud all along the bottom; then a layer of lime and dung, and mould at top; others will lay some mould first at bottom, and then a long row of stone-lime, and on both sides of it, mould and dung mixed, which is all, some time after, to be mixed with a spade or shovel, till it is incorporated into one fine body, and then it will fertilize almost all sorts of land, as being an agreeable manure, which will rather
destroy

destroy than breed weeds: but of lime and these, more hereafter.

By mixing your dung, mould, and lime together, you will bring all the compost into a fermentation, which, by the help of the dung and lime, will burn up all the seeds of weeds, and the small roots that may be contained in the mould; the whole made to run into a fine body. If you are to mix lime with only pond, river, or ditch mud, then put one load, or five or six quarters of lime, to ten large loads of mud; observe also, that this must be once turned at least before you lay it on for wheat, that all may be duly incorporated and made fine. There are some farmers who use lime in a dunghill, thus: first they lay a bottom of horse, cow, ass, or hog dung, of two feet thick, upon which they spread a covering of earth two feet thick likewise; on this they lay what lime they think fit, then dung, then earth, and then lime as before, and so on, till they have their quantity, and at last cover with turf or mould, to keep the sun from drying it too much, letting such a heap lie rather broad than high, that they may the better ferment together; and this they never fail to turn once at least. Lime is so great a shortener of mould, that, if it is thrown but thinly on ploughed land, in ridges especially, in *September*, *October*, or other winter months, it will wash

into,

into, and fine it, with the help of frosts. Lime alone is used for *French* wheat, as well as common wheat, and is also excellent for meadowground, turnips, pease, &c. as being a cool, sweet, rich dressing, and helps the corn to kern; hollows the ground, sweetens the bite of grass, and in many places is made to supply dungs and other dressings; but more of this in proper months.

CHAP. V.
OF CHALK.

CHALK is a mineral that is of most exquisite service in farming, as being the greatest alterative of any other to our clays, loams and gravels: without this great improver, the many fine crops of corn and artificial grasses could not be obtained that are, because it cures the clays of their sour, austere, cold, hard and tough qualities, and establishes for sometimes twelve, or twenty years together a lightness, warmness, sweetness and shortness in their room; and so also in the loams and gravels, this earth does wonders by converting their sour and binding natures into their contrary qualities, whereby the plough performs its operations more easily, and the ground becomes more fertile, by causing these earths to emit their falts to their feveral vegetables with freedom, which otherways would be tenacioufly fixed and kept from nourifhing the corn and grafs, that would then want their vital affiftance; fo that it evidently appears, that thofe who chalk moft, receive the greateft crops: but in the vale they ftand not in fo much need of this moft kind earth, becaufe their black and bluifh marly-clays are naturally fhort in themfelves, and readily yield to the power of the air and rain: whereas our red clays in the chiltern are quite different to thofe of the vale, and require the help of chalk or fand to reduce their furly bodies, as they will not be prolific in their productions.

Where the chalk may be moft commodioufly drawn, is generally in the middle of a ploughed field: this is according to the late and prefent practice, by the moft judicious of our farmers, and for fo doing they affign this reafon; becaufe when the chalk is drawn, a great hole or pit is evidently caufed, and by being in the center of the field, the plough, by traverfing the ground on all fides, in time will bring down and drive the adjacent earth into it, and fo by degrees will fill it up in fuch a manner as to make the place little or nothing the worfe for either fowing or ploughing. In many places thefe holes or pits are made in an angle, or on a grafs baulk of

the

Chap. V.   OF CHALK.

the field, and then there is never any chance of filling them up, otherways than by the great trouble and charge of bringing other ground and casting it into them, as it is likely to be my case in a field that I bought amongst some others about five years ago, where I found a large, deep pit made close to a hedge, which had once like to have occafioned me the lofs of a horfe.

But where a wood or fpring adjoins to the field, or near it, then this fituation may be moft proper for the finking a pit; or if it is made clofe to the wood, it will have this conveniency, that it will be eafier fenced againft cattles falling into; and trees or other wood will neverthelefs get up in a little time, fpontaneoufly from the fibres of the adjacent roots, and then grow up and run fafter than ordinary in this hollow cavity of ground, as is often feen in large trees that grow in dells and holes of the earth, where the fhade and water have more power in the nourifhment of all vegetables than on plain ground.

Sometimes I have known the men make more than one pit in a field; to find in a fecond better chalk than they did in the firft, where it prefented it felf good for a little way, till a vein or cruft of the hurlucky, ftony fort prevented any further penetration, which caufed the fecond attempt at fome diftance; and anfwered the owner's satisfaction; for here appeared a fat, soft chalk, with a yellow coat or covering, and this is a certain indication of its goodness; where out of this pit, two adjoining fields were chalked just by my house.

The chalk drawer finds a wheel rope-barrow, and all other tackle, and also sinks the pit for the price of eight pence a load, each load containing twenty wheel-barrows full, which they also for that money spread all about the field. Twenty five or thirty load will well chalk an acre of ground, which by discreet ploughings will last twenty years. But here I must stop my pen to expose the inconsiderateness and folly of all those who by thinking to save charge, oblige the chalk drawers to put on six acres of ground, no more chalk than would thoroughly dress one: this in proportion I have known done, where it could be afforded to full dress. This defective management causes many fatal mistakes, for when ground is so chalked, the plougher and sower are apt to order their matters as if the land was full dressed, which often deceives the owner; for to cross-crop this ground, or sow this ground as if it had its due quantity of chalk, is wrong, and will force it to complain in a little time of such hard usage. This work must be sure to be done about *Michaelmas* or a month after, that the frosts

Chap. V. OF CHALK.

frosts may shatter and crumble the chalk all the winter as it lies on the surface of the earth; otherwise if drawn in summer it will grow hard and petrify by the sun's great heat and the dryness of the air; or if ploughed into the earth in lumps, it will so remain many years: as a farmer at *North Church* near me did, who, by mistake, drawed his chalk in the spring, and ploughed it into the ground in lumps, that still remains so, though it is several years since he had it dug.

This chalk when thoroughly reduced into a powder by the winter frost, is called the best of dressings; not as it is rich in itself, but as it sweetens, shortens and dries the clay's body, and so makes it fit to receive, and easier join and mix itself with other manures, that may be thrown upon or ploughed into it; whereby it becomes a loose earth, and lets the waters in the winter and cold springs through its pores, which before used to hold them, and thereby chilled and starved the corn. When therefore this chalk is so reduced by the weather, at *Candlemas* plough it with or without a fin on the share, very thin into the ground, and by this one ploughing may be sown beans, either strained into the thoroughs and afterwards harrowed down, or else by sowing the beans all over the field first, and then plough them in very shallow, or to sow them

half

half under and half over, thorough and harrow well; and so after this manner may peas be sown.

One of our best farmers this last spring, eat off his turnips early, and chalked his ground well; then the beginning of *March* he gave it one ploughing in broad lands, and harrowed in his barley and footed it on the top. This is an excellent way to lose no time, and hereby the grain has no less than three dressings. If sands are to be laid on this red clay, then twenty or thirty cart loads on an acre can't be too much for this purpose, to shorten so tough an earth as some of the strongest sort is.

But to spoil a serviceable level surface, by digging into it and making and leaving pits open, may be of very ill consequence; and, to prevent mischief from open pits, the new invention took its rise of digging only a round hole of three feet diameter, enough for a tub to be let down, and drawn up with chalk, for chalking our *Hertfordshire* ploughed fields; and this hole we now commonly make in a hedge, on purpose that it may be the more out of the cattle's way, to prevent any danger of their falling into it. Whereas, heretofore, we used to make and leave large pits, perhaps, fifteen, twenty, or more feet wide to come at our chalk, and this in the middle, or some other open part of a field, which has occasioned many misfortunes.

CHAP. VI.
OF MARLE.

THERE are four several sorts of marle, *viz.* the fustian, the cowshit, the black steel, and the shale: the fustian sort is an earth composed of a fat loam and sand of a reddish colour so soft and loose that they spit it with a spade, and lasts but four or five years, though relieved with other dressings in that time: the cowshit, which is the richest sort, looks to be an earth mingled with lime leaving many little white specks in it, and will last seven years with assistance; this is all spitted or thrown out with a spade: the black steel marle is of so hard a nature, that they dig it with mattock and spade, and will not all dissolve in seven years, nor will all its goodness be spent in twelve: the shale marle is of all colours and of a stony nature, which obliges them to peck and hew it, and then it comes like bits of stone; yet this will last but four or five years though helped with other dressings: of these they sometimes lay above 500 heaps on an acre; and after it is spread they let it lie dissolving the remaining part of the summer and the whole winter till they give it one ploughing between *Michaelmas* and *Candlemas*, and afterwards harrow in oats.

A gentleman dressed a three-acres piece of light gravelly sandy ground with three hundred load of marle, but two years ago; and since that dunged it well all over. This double dressing must surely be one of the highest improvements of such ground, for nothing agrees better with a gravelly sand than marle, because its short dry parts are toughened by the marle, and brought into a much moister condition than it was in before. Now these two qualities being contrary to the nature of gravel and sand, this mixture reduces it into a loamy body, and thereby enables it to make a longer lodgment of all dungs and manures that shall be incorporated with it; for while this soil remained in its original, hungry, short, loose condition, dungs and manures were soon eat up by it or soon washed away, because the particles of gravel and sand are generally of a globular make, and therefore cannot lie so close together as those of loams do. Hence it is, that marles, or rich clays, are as natural a mixture with sand, as sand is with clay; because these bring their opposite soil into a medium earth; an earth that may be justly called a loam, and a loam the best of earths, as being the most general sort of all others, for nourishing almost any vegetable. On this account, those gentlemen who are possessed of marle-pits have just reason to think themselves happy, which several coun-

ties

ties in *England* are strangers to. I do not know of one pit of this rich earth in *Hertfordshire.* But, in my late travels, I have seen several of different colours, and all of them of a most fertilizing nature. To supply which defect, we are obliged to be at the great expence of buying and fetching at many miles distance, sheeps trotters, cows and oxens hoofs, hogs and cows hair, oil-cake powder, malt dust, lime, soot, ashes, rags, pigeons, hens, and rabbits dung, &c. And thus we are forced to lay out, in advance, the value of an ordinary crop of grain, in hopes of obtaining a very profitable one. Though sometimes we miss of our aim, and fall short of our expectation in this matter; for when a dry hot summer happens, our grain on gravels, chalks, and sands, commonly fare the worse for some of them; for, by their hot nature, they help to impede, rather than forward, the growth of the crop.

CHAP. VII.
OF LIME.

LIME, in different places, is attended with different management. In *Surry*, they lay one bushel on a heap taken out of a low cart, and this at every pole-distance throughout a field; and, when slaked by the weather, with a shovel, a shovel, they throw it over the land, plough it in, and then directly harrow in their wheat-seed on the same. Or, when the ground is ploughed and harrowed, the last time but one, the lime may be sown and ploughed in with wheat-seed in two-bout lands: or, when the lime is sown in a little quantity, as that of sixty bushels on one acre, when it is flaked, the lime and seed may be ploughed in together in broad-lands, and it will nourish two crops well.

But, in some parts of this country, they lay on a whole kiln of lime on three acres, containing nine loads, and each load forty bushels; which, with some alternate dressings of dung, or other sorts, will last ten years; for this sort of manure greatly hollows the ground, causes wheat to kern, sweetens the land, and kills weeds and insects.

On the fourth day of *August*, 1738, I saw a large field just ploughed up and rolled in *Surry*; and then half a bushel of lime was laid, at every pole-distance, in rows throughout the same, for sowing wheat. Some let lime lie, till it is flaked, before they throw it about with a spade or shovel.

In some parts of *Essex*, as well as in *Surry*, *Kent*, and many other places, they give two or three ploughings to an oat-stubble, and then lay one bushel of lime on every rod of ground, where, after a very few days, it commonly flakes;

then

Chap. VII. OF LIME.

then they plough shallow, and harrow in wheat in broad-lands; and thus they say, that, with a little other dressings afterwards, lime mends lands for seven years. In some parts of *Surry*, when they lay a bushel of lime in a heap in the field, it is their way to throw a little mould over it, that it may the more gradually and leisurely slaken, and by this means, it will swell to a great degree; then they spread it over the land with a shovel, and plough and sow rye, wheat, or other grain.

If lime is made use of as dressing for wheat, it must be slaked, and then immediately sown hot on the last ploughing, all over the ground, about 25 or 30 bushels on an acre; this should lie 6, or 10 days, and then plough and sow the wheat in stitches as at other times, which will secure it against cold, wet weather in winter and spring, and make it look of a deep, dark green, when the neighbours will die and look yellow by the chill of frosts, cold and wet.

Lime must lie, and as soon as slaked, sown over the ground, to the quantity of twenty, thirty, or forty bushels on an acre; immediately after this, the turnip seed must be sown, and both harrowed in together.

CHAP. VIII.
OF BURNT CLAY.

MOLE, or Ant-hills, confifting of a reddifh, clayey nature, with fome mixture of loam, that had been cut up with the iron-plated fhovel, or fpade, and carried to one great heap, where they lay till *May*, or *June*, and then were burnt all in one heap into afhes; thus: firft, the farmer begun to lay a little faggot-wood for a foundation, and on moft of that he laid on fome intire loam or turf, and on that, fome of the ant-hill earth; then he began to light his fire, and, as it burnt, he put on more, till he increafed the heap to a very great bulk, which to do in the beft manner, the fire-man continued laying on the ant-hill earth, fo carefully, as to keep in the fire very clofe, for, if it had burfted out, it would have done but little fervice.

Thus, when all was burnt and calcined into afhes, he laid them on his meadow-ground, but they did not anfwer his expectation; becaufe a great deal of this reddifh clayey earth remained, after firing, in the hard body of many pieces, almoft like bits of brickbats, which he was obliged afterwards to collect and carry off, as being too hard a fubftance for the weather to diffolve: this proves, that burning clay in a heap, in the open field, is not fo good a way as burning it in a

clamp;

Chap. VIII. OF BURNING CLAY.

clamp; and therefore a late author, in his treatise of improving clay-grounds, is certainly absolutely right, in his publication of a method how to burn wet clay in the greatest perfection in a clamp; because, by this method, clay is brought to yield more salts, than when burnt dry, and also burnt into a perfect ash, or powder.

Now why the clay ant-hills were burnt to ashes, and not served as those in the vale were, I think wants no explanation; and therefore I have no more to say on that account; but I think it one of the greatest improvements in husbandry, to burn clay (the red sort especially) to ashes, for manuring either meadow, or ploughed-ground, provided the clay can be dug without any damage. I confess, I have burnt several loads of red clay into ashes in a heap in the field, but never in a clamp, because my quantity of clay did not make it worth my while, for it was only what I had dug out of a pond.

It is not only to burn clay into ashes, but their best application is also to be endeavoured. It is certain, that the salts of all ashes whatsoever are a most powerful enemy to the breed, and damage of all insects: and, therefore, it is observed by several authors, they are good to sow over turnip-ground, &c. to prevent the fly and slug, and so they are. But clay-ashes are not an infallible remedy; for, by great and continued rains

rains, their salts will be washed away into the earth, so as to make them lose their efficacy in keeping off insects.

However, they are very valuable, as they serve to nourish the turnip-crop. And, for doing that, these and all other ashes ought to be sown on the top of the young turnips, as soon as they appear in their first green leaf; and then two cart-loads of them will be better sown than one on each acre, both of which will contain sixty single bushels.

But I insist on it, that neither these clay-ashes, lime, nor tobacco-dust, will assuredly prevent the insect, because their powdered natures may be rendered ineffectual by rains for this purpose. Clay-ashes are a most powerful nourishment to all wheat, oats, barley, pease, and beans, if strewed or sowed by the hand out of a seed-cott over them in the spring-time, to the quantity of three cart-loads on one acre, and the same quantity on every acre of meadow-ground; then they will prevent the growth of moss, bring up the honeysuckle, kill all rushy, sedgy grass, and kill or prevent the damage of the grub, or cankerworm.

There is more than one way to burn clay into ashes; one is, by burning it in a heap in the open air, as we do peat, with the help of roots, faggots, or other offal-wood and some turf

Chap. VIII. OF BURNING CLAY.

or mould put next on it, to kindle and light the fire, till it burns into the clay; and, then only with a gradual increafe of fire, and cover of dry or wet clay, or both, on the outfide of the heap to enlarge the quantity of the clay and keep the increafed fire at the fame time from burfting out, the workman may carry on his burning with fafety till he has burnt hundreds of loads of clay in fuch a heap.

I fhall not fo much enlarge on it as otherwife I would do, but obferve, that the burning of clay in a clamp is, by two authors, recommended as the beft way of all others; and that the burning of clay wet is much better than when burnt dry, for reafons the author on improving clay-ground affigns; but, for my part, I know of none, as yet, that burns clay for manure in a brick-clamp, and therefore fhall here only remark, that, when clay is burnt to afhes, fuch afhes are not to be carried into the field as they are, and as fome have done, and laid them in a promifcuous condition on the ground; no, this would be perfectly wrong management, and have the fame chargeable effect, as happened to two feveral gentlemen through the ignorance of their bailiffs, who carried their clayafhes, and fpread them over the grafs-ground they were to improve, as they were burnt, without firft feparating the grofs fort from the fine; and thus

pieces

pieces of burnt clay, like little pieces of brickbats, were laid over the grass-ground, in assurance that the frosts would shoal and dissolve them; but, on this account, their expectation was deceived; for these pieces of burnt clay had too much of the clinker nature in them, to dissolve by weather, at least, not in a very little time; which obliged the owners to be at the charge of having them picked off the grass and carried away. A new method of applying these ashes in a far more profitable manner than hitherto has been done by any person I know besides one particular gentleman's bailiff, who most commonly every year burns more or less of these ashes; for to burn clay into ashes, and not know how rightly to apply them, is but doing the work in part.

January is the time for sowing clay-ashes on meadow-ground. And, if such a ground is overrun with moss, four or five inches in heighth, it will be eaten or burnt off by repeated dressings of ashes, as I have been an eye-witness of; and where moss has got possession of meadow-land, it certainly concerns the owner to get it eaten off by ashes with all expedition, because this has roots, and is nourished by the earth, to the impoverishment of the grass. I knew a meadow of near five acres, that was so over-run with moss, that it returned the owner but half one cart-

cart-load of hay in all one summer; when, if it had been in good order, it would have yielded ten loads at least, as it has done; since it has been cured of this destructive weed; for this was an upland meadow of an extraordinary rich soil.

CHAP. IX.
OF PEAT.

PEAT is taken out of a black moorish ground at *Newbury*, by a wooden, narrow scoop, which brings it out like a long narrow brick; this they lay on the ground to dry in the summer time, and then sell it for eight shillings a waggon load as provisional fuel for families: but when it is to be used for a manure, after it is dried, they burn it in heaps of ten, twenty, or thirty loads, laying on more peat on the outsides, as the fire increases within, to keep it from having too much vent: however, in time there will appear a considerable smoke; and it was on the twenty-third of *May*, 1737, that I saw about ten great heaps burning for this purpose, near where it was dug. The great use of these ashes was found out about thirty years ago; but in a little time after were brought into disreputation, by their imprudently laying on too many at a time; which burnt up the corn.

Afterwards they found that fix or ten bufhels were fufficient to be fown over an acre of wheat, peafe, turnips, clover, rape-feed, or fainfoine, as early as they conveniently could. But, as I faid before, they are afraid to fow it over barley, left a dry time fhould enfue and burn it up; for thefe afhes are reckoned to contain three times as much fulphur in them as in the coal-afhes; and this they reafonably imagine from their great brimftony fmell, fparkling and jumping, when they are ftirred as they are burning, and drying up the corn by their too great heat.

Thefe peat-afhes, and likewife thofe from wood or coal, will help to keep off the flug from peafe and other grains, by the falt and fulphur contained in them, and very much conduce to their prefervation in cold wet feafons. But there is no fuch danger to be feared from the afhes of that peat, which grows as a turf over fandy bottoms, as great quantities do on *Leighton-Heath* in *Bedfordfhire*, for thefe are as much too lean, as the other are too rank.

About *Newbury*, in *Berkfhire*, I think I faw the greateft peat-ground in *England*, I mean, where they burn the moft peat in heaps abroad, purely for making afhes to drefs land with; and here it is furprifing to fee fuch numbers of large trees taken out of the ground, that lay buried eight or ten feet deep, fome retaining their natural fubftance,

Chap. IX. OF PEAT. 67

substance, and others decayed, and as rotten as touchwood.

Oak, deal, and other timber-trees of a prodigious size have been found so sound, as to be made use of in building houses. Stags horns and many other things have been likewise discovered ten feet deep. How these should be here has employed the thoughts of many, whose opinions are various: some conjecture, that at the deluge, when the waters covered the face of the whole earth, and all nature suffered, not only the animal, but the vegetable part also had its share, when every high hill under heaven was covered with fifteen cubits depth of water; and by the waters prevailing and continuing on the earth one hundred and fifty days, probably (say they) it might make such devastation among trees, that they might be torn up, and, by the rapidity of them, be brought to these vale, or flatlands, where, as the waters abated, the trees were left, and, by the weight of their bodies, sunk to their center; for it is observed, that, the bigger the body, the lower the tree descended.

Others are of opinion, that some great floods and storms of winds of later date may be the cause of this wonderful accident. But certain it is, that many trees of a large bulk, unexposed to either wind or water, lie here entirely wasted to rottenness, and are as soft as butter.

F 2 In

In the next place, I shall proceed to give an account of the nature of this peat-earth. As I observed before, peat is commonly found in flat grounds, but, as to its depth of lying, it is uncertain. In some places, the bed, or *stratum* of peat, is found six feet under the surface of the earth, and, to come at it, they are sometimes obliged to dig through one foot of top black mould, or loam, and, after that, through a white maum three feet thick. Sometimes a gravel lies betwixt the surface and the peat; but to know the true peat it is very easy, by its black colour, its hollow light body, and its being full of mossy fibres, or thready roots, and not having any other earth intermixed with it; not but that there are divers kinds of peat of various colours, but the best is the jet-black sort, which generally lies in bottoms next to rivers.

In the fen-countries of *Cambridgeshire*, where the waters lie on the peat-grounds great part of the year, I am of opinion they are, in this manner, the cause that peat is very light and hollow, and by this the virtue of their peat and its ashes is much lessened, and its ashes fewer in quantity. On the contrary, where peat lies wettish, but much drier, it has generally a more compact and closer body; and, the heavier such peat is, the stronger and heavier the ashes are,
and

and the more service they will do, where-ever employed.

When I was at *Bristol*, in the year 1737, the people told me, that they were ignorant of the virtue of soot, as it related to the manuring and fertilizing land, insomuch that they threw it away on common street dunghills. These peat-ashes were formerly served in the same manner, until a farmer, whose genius aspired to further improvements in husbandry, than were in common practice, ventured to sow these ashes on his ground; and, finding a surprizing effect of their goodness, continued their use for several years, before their value was publickly known, to his great profit.

Between *Hempstead* and *Watford*, in *Hertfordshire*, about the year 1738, a low meadow, that lies contiguous to the river's side, was broke up, and a most excellent peat discovered, not inferior (as is reported) to that of *Newbury* in *Berkshire*, because eight bushels of these, which are equal to twelve common *Winchester* bushels, are sufficient to manure one acre of corn-ground with; and all that quantity costs but five shillings and four-pence, at eight pence a bushel, and it is said to do as much service as twenty bushels of soot. A cheap dressing indeed, thus to return great crops, if the season is kind, both of corn and grass, besides preventing the damage of insects.

A peat-ground, near *Langley* and *Hempstead*, has not been broken up for this purpose above four years, occasioned first by the owner's taking a survey of that at *Newbury*, and the information he received there of the great service and value their peat-ashes were of: On this he purchased three or four meadows lying near a river's side, which produce the peat I have been describing, and seems to be as good as that of *Newbury*. Here Mr. *Lea* proposes to furnish any house with peat, to burn it as fewel all the year in grates, for fifty shillings, provided they take care of the ashes, by burning no wood or other fewel with it, and let him have them all neat; for, if these are all saved and sifted, they will be of a whitish colour and very fine, and in goodness near, if not quite as good, as those burnt in heaps in the meadow.

Another sort of peat is called ling-peat, such as the common people pare off the surface of dry commons, as that is near *Leighton* in *Bedfordshire*, and many other places in *England*; but this produces very poor ashes, because the ling or peat comes off a poor soil, and therefore is thrown to the street dung-hill; yet at a place about two or three miles distance from that, in a low meadow, not near a river, there a peat is dug called bog-peat, and is like that at *Newbury* and *Langley*, lies deep in the ground as they do, and of the same colour and goodness. Also about two years

years ago, I am told, a certain gentleman in *Bedfordshire* being informed of the service of the *Langley* peat-ashes, got hands from thence, and fell to work; and, having discovered an excellent sort in his own estate, refused to sell any, because he will keep all he makes, for his own and his tenants uses.

At *West-Hyde*, they say, that *Newbury* ashes are so full of sulphur, that they dare not sow them on wheat in the quantity they are sowed on pease, or artificial grass, lest it cause them to grow too rank: But of late they have ventured to sow them in a lesser quantity, as seven bushels instead of ten, on the wheat crop. Also, of late, they say, peat is burnt at *Newbury* in a clamp, like a brick-kiln near *London*; wherein, like that, they leave places for the fire to go by flues from one part to another, and secure all the outsides of it very close, to hinder any fire coming out. *Newbury* ashes were made a trial of at *Taplow*, between *Rickmansworth* and *Uxbridge*, thus: —Nothing was sown on one part of the field, the other part was sown with smith's ashes, and a third with these peat-ashes: The smith's coal-ashes did good, but the peat-ashes exceeded the smith's, as much as the smith's exceeded that part which had no dressing on it.

Near *Langley* they do not burn peat in the manner of a clamp, or brick-kiln, but only in heaps of one or two hundred, or more loads in a heap; and this work is carried on, almost all the summer long, in an open meadow, just by where the peat is dug. It is dug, or scooped out, in narrow pieces, near two feet long, in shape like a brick, and carried directly to the heap intended to be burnt; where, with a few faggots, a heap is soon set on fire, that must be kept lined, or covered without-side, according to discretion, with more peat, so that the fire must be neither suffocated, nor have too much vent; for so prompt is peat to take fire, as being a spungy fat earth, that a great heap need not be long attended, and therefore one man can manage the fires of several, at one and the same time, by reason they will gradually burn and calcine almost of themselves, into a reddish coarse sand, like heavy ashes.

Accordingly some heaps that have been little regarded after taking fire, have burnt, little or more, for two months together. And why these ashes are of a reddish colour, and more coarse than those made from peat burnt in kitchen-grates, is, because here the fire is confined under cover, and the smoak very much prevented evaporating, which in kitchen grates have both a greater liberty of a more expeditious and free consumption, and therefore the ashes are burnt whiter; for it is the nature of smoak to tincture all things of a very brown or reddish colour, that

in

in a moderate degree are confined to it; and black if they lie very near, and long by it.

Then, after the peat is burnt, and calcined into these ashes in the meadow, some are laid under cover, to be kept from the wash of rains; others are laid up in the open meadow, in great long heaps like a hog's back, which, by their close lying, and ridge shape, will remain very secure from damage all the winter, and in *January* or *February* next, they are brought under cover to be sifted, and sold to the farmers and gardeners; for, as they are burnt in large heaps, there will be great quantities of hard bits, and pieces of burnt earth, that must be first separated. Likewise those that are made by peat burnt in grates must be also sifted, for the finer the ashes are made, the further they will go, and do the more good.

It is certain, that if these ashes after sowing are attended by a long succession of dry weather, they must not be expected to do much good the first year; nay, sometimes they will do more harm than good, by assisting dry hot weather to scorch up the corn or grass. But then this is no more than what coal-soot will do, that we give one shilling for every single *Winchester* bushel delivered at *Gaddesden*, twenty eight miles from *London*: But, if showers fall in time, then their profitable effects may be soon seen in perfection. Peat also, especially when it is burnt in grates

not

not thoroughly dried, will yield an offensive smell to the victuals that are dressed by it, and to the company that sit by its fire: because this light, spungy, subterraneous, black earth, being full of mossy fibres, or roots, casts out such a brimstone smell, as makes it disagreeable to all within its reach.

CHAP. X.

OF COAL-ASHES.

THESE excellent ashes are much used by gentlemen and farmers who live within thirty miles of *London*. By gentlemen chiefly, for improving their meadow-ground: by the farmer for improving his meadow-ground, his artificial grass, his wheat, and his barley crops, &c. First, the farmer, for his meadow-land, thinks it a cheap manure to lay forty bushels of these coal-ashes on each acre, for producing plentiful and early crops of the best of grass, for three or four years together. Secondly, The farmer thinks it of the like service to lay the same quantity of these coal-ashes on each acre of his clover, ray-grass, trefoile, lucerne, or sainfoine in *March*, as the best time of all others, and that in the beginning of it, because these coal, like peat-ashes, are with difficulty forced to part with their sulphureous quality, which is

the

Chap. X. OF COAL-ASHES.

the riches of them, and therefore require great and long washings of rain to make them part with it. On this account it is, that these ashes become more and more in use, especially for thus sowing them on artificial grass, as being thought by many to be more profitable for this purpose than soot, because they are of opinion, that as these coal-ashes cost but three-pence or four-pence a bushel, and soot sometimes ten-pence or a shilling a bushel, and very much adulterated into the bargain, they are of the greatest profit for artificial or meadow grass. They are also much used in making a compost of dung, or highway or pond mud, with chalk, or lime, or soap-ashes, for meadow-ground; and some sow coal-ashes naked, as they come from *London*, on their growing corn in *February* and *March*, to great advantage, particularly when a wet summer follows their application.

There are many farmers in *Hertfordshire*, that grudge not to give three-pence or four-pence a bushel, or twenty-pence for a sackful of them. They, of late, have got into such reputation for fertilizing natural and artificial grasses, that great quantities are every year made use for this very purpose; but few farmers believe they will do service to ploughed grounds. Others say they will do more good on ploughed ground than on meadow, that is, on ploughed ground, that clover, sainfoine, trefoile,

and

and ray-grafs are fown on, becaufe ploughed ground will fooner draw them into it, than the hard cruft of meadow ground can. Accordingly, fome think likewife, they will prove efficacious on wheat if fown early in this month. Next to foot, coal-afhes are preferred, as a choice manure for clover, fainfoine, and all artificial and natural graffes to that degree, that many farmers believe thefe afhes are better than foot for all forts of graffes. But this I cannot believe; however, if coal-afhes are made from private fires, and kept under cover till they are thus ufed, they will do prodigious fervice for two or three years together, if a hundred bufhels of them are fown over one acre and a half of fuch grafs ground. About *Rickmanfworth*, the farmers fow thefe afhes on their firft crop of clover that grew among barley, and they will foon make it come up in their gravelly loams into a fine head, for giving their fuckling ewes a bite as foon as the rye they fowed for them laft year is eat off; which, with only the fheeps dung and ftale, will caufe the clover to laft two or three years in good heart, without any further affiftance from dung or manure.

On the clay land, that has been fallowed in broad-lands, lay in heaps, on one acre, the quantity of fixty or a hundred bufhels of coal-afhes, the more the better, and fpread them with a fhovel or fpade in a dewy morning to prevent

prevent their flying too far, and then plough them in as shallow as possible into broad-lands again, across the last way, or into four thoroughed stitches, or by hacking the ground, this will reduce it to a loam.

CHAP. XI.
OF WOOD-ASHES.

WOOD-ASHES will do great service both to natural and artificial grasses; but they will not last above half the time of coal-ashes, because they are of a much softer and looser nature than the coal sort are; and, therefore, if wood-ashes are laid on at *Lady-day* it is time enough, for these are quickly washed in, and therefore will last but one year well. One hundred and sixty bushels of wood-ashes are equal, for this purpose, to half the quantity of coal-ashes; and either of these two last quantities is but sufficient to dress one acre well of grass ground.

I have sown five cart loads of wood-ashes on one acre of meadow, and found them not too much, provided they are sowed on the same in *December*.

There is a new piece of good husbandry of late acted by some that buy their soot in summer, and lay it in the same field that is to be sown

sown with it, where they thatch the heaps with straw, that keeps off all weather.

Ashes, either of the wood or coal sort, are a very good dressing for corn or grass grounds; the first we buy for three halfpence the single bushel, and sow it out of a seed-cot in our wheat, rye, or barley-ground, in *February* or *March*, by throwing twenty, thirty, or forty bushels on an acre of the surface; but the coal-ashes are so full of sulphur, that eight sacks, or thirty-two bushels, are enough for an acre on some soils, and are generally used on grass grounds, where they will burn off the moss, and suddenly breed the honey-suckle grass to that degree, that I must needs say I think it an excellent manure, especially if rains soon follow their sowing. Farmer *Wright* of *Barley-end*, this last spring, put on about twenty-four bushels on an acre, and had such a burthen as he never received in one year on the same ground these thirty years past.

It was the expression of a judicious man I was in company with, that a bushel of coal-ashes was as good for sward ground as a bushel of our wood-soot that is here sold for six-pence; because the sulphur in one was so much hotter and more preferable to the salt in the other, and, next to soot, is one of the best remedies for killing the clob-weed, rush, and other coarse grasses that too often infest the meadow-grounds.

The

Chap. XI. OF WOOD-ASHES.

The wood and coal ashes of late have gained great reputation for their many fertile uses, as well by themselves, as mixed with short horse-dung, cows, fowls, and other dungs and stover; so likewise have those made from burnt clay, mould, peat, turf and ant-hills: yet I know of a large meadow containing two or three hundred acres, that were run over with ant-hills very thick; the hills were cut up with the wide sharr-plow, and the grass-part put into heaps with the forks to lie and rot, but the mould-part was immediately scattered over the ground with the shovel; the turfs being not to be meddled with till they are got fine by lying, fermenting and putrifying almost a year, and then it is to be spread about the field: The reason for not burning the turf into ashes, was, because in this manner it contained and kept in its salts better, than if often turned and exposed to the washings of great rains, or burnt; and also because this ground lying low and wettish, this way would better thicken it, and keep it drier hereafter, than if it was reduced to ashes, which I think is very consonant to good husbandry.

But whoever buys these *London* ashes, must take particular care that they be pure and not adulterated with dust and other trumpery, for then they will likely deceive your expectation; and so will the ashes from great brewhouses, smiths forges, glass-houses, and other places in and about *London*, where by the fury of their large fires the ashes are burnt to that degree that little goodness is left in them, which causes those from private families to be more excellent, as they are less burnt, for promoting the growth of vegetables.

In the country we make a particular distinction between ashes from wood (the hard sort especially) and those from furze, fern and straw, for the latter are not so valuable as the former, because the wood sort exceeds those from furze, as much as the furze does those from fern, and fern from straw. Yet notwithstanding the great goodness there is in these soap-ashes and all others, they may be used to a fault; for I have known this dressing so often repeated on one and the same piece of ground, that they have lain so thick at bottom as to choak the growth of the grass to its very great damage.

However, to do justice to this fertile dressing, I must commend it next to soot, lime, and fowls-dung, for not breeding weeds, at the same time it nourishes the earth, as the horse and some other dungs are apt to do; which has raised my esteem so much for the

London

London ashes, that when I have sent stack-wood thither, my team has brought back a load of these from the town lay-stalls, for which we pay twelve-pence there; but these are wrong places to take them from, because as they here lie open to the weather, they are often washed by the rains, and so lose a great deal of their sulphur; which is the essential part of their dressing; whereas if the ashes were saved under cover, as they are brought out of the houses, their goodness might better be depended on.

CHAPTER XII.

OF SOOT.

SECT. I.

Soot on Wheat.

MARCH is the proper time to lay soot on wheat, because by this time, we reckon, snows do not lie long on the ground; for deep snows are oftentimes a great enemy to this manure, especially when they are dissolved by sudden cold thaws; for, then, new-sown soot is apt to be washed into the ground too soon by them, and leave the wheat but little the better for the great expence a farmer is at, of laying out twenty shillings in soot to dress an acre. Let twenty bushels of coal-soot be sown over each acre of land, as thin as can well be sown out of a man's hand, over wheat sown on broad-lands, or in two-bout-lands; and then, if the soot lies but two days, free of the wash of rains, it will take the ground so gradually, as to prove one of the best of dressings, by enabling this golden grain to withstand the severity of frosts and winds, that commonly reign in the spring-season, and also bring the wheat under such a quick growth and cover, that the droughts cannot likewise be of prejudice to it. Thus this noble sulphureous manure enriches the same ground for three years successively, and is so efficacious, that we seldom fail of having a good pea-crop the next year after.

Soot is a most powerful manure, made use of by many of our best farmers in the *Chiltern*, though but by few in the *Vale*, because they grudge the charge, and supply it with their flocks of sheep; not but that it will do as much good there as any where: its chief use is on wheat, rye and barley, and is commonly sown from *Christmas* to *April*, on any of them out of a seed-cot as thin as can well be, to the quantity of twenty or twenty five bushels on an acre, for if the soot was to be sown sooner, the winter rains and snows would endanger its goodness being washed away too early; but of late some

sow

Chap. XII.   OF SOOT.

sow their wheat and rye in *August*, that they may eat their first head in *November* with their sheep, as reckoning it to be near as good a crop as of grass, and then will dress them with soot about *Christmas*, that they may the sooner get a second head.

But when they sow soot, in the vale, it is done about *Candlemas*, and then they sow eight bushels on each half acre land, as thin as possible, so they cover all the ground; and, if a dripping-time follows, it does a great deal of service, in keeping off chills, by warming the roots of the wheat; and, indeed, if no rain happens for some time, the very moisture of this sort of ground will draw in the quintessence of the soot to its great benefit.

The same quantity will do on ploughed-ground, sowed over wheat or rye, in *January*, either on broad-lands, or on two-bout ridges. But, I am of opinion, it is better let alone, till the latter part of next month, for then you will be more secure against the fall of deep snows, which, in *January*, are very apt, by a sudden thaw, to wash away the soot too soon; for many farmers have lost the benefit of soot this way, and by very great and long rains that have fell presently after the soot has been sown, and washed away its goodness before it could regularly take the ground, which it should have some dry days to do, to make it truly efficacious.

SECT. II.

Soot on Barley.

IF we dress barley with soot, about a week or a fortnight after it has been sown, we sow twenty bushels of it over each acre; but others think it a better way to sow this quantity of soot, as soon as the barley-seed is sown, and harrow both in together; and indeed, I have found it the best way, by reason it lies in this form of sowing, closer to the kernels, than when the barley is sown first, and harrowed in, and the soot sown afterwards; nor has the soot so much power this way to burn the blades of the barley, in case a dry hot season should follow, as it has when sown after the barley is harrowed in; and so efficacious is such a dressing of soot to the ground, that a pea-crop sowed on the same, in the next spring season, is hoped for in a plentiful manner, and by this means seldom fails the owner's expectation.

But the manure of soot will not answer to profit, if made use of in three sorts of land; in a loose light sand, in marsh lands, or in three or four-bout lands. In the first, the earth is too hollow and spungy to retain it long enough on its surface to assist the roots of the barley; in the second, the land lies too flat and wet, to give the same benefit to the grain; in the third, the wa-

Chap. XII. OF SOOT.

ter-thoroughs are too deep, and the ridges too narrow, to receive a due affiftance from fo light a body as foot is; and, when it is wafhed off into thefe thoroughs, it is in a manner loft, becaufe little or no corn grows here; but, when it is fown on barley, that is fown on broader lands, it will anfwer, as I have before obferved. Secondly, I know one, and only one *Vale* farmer manure his barley-crop with foot, and, I was a witnefs, it did great fervice, though not fo much as if he had made ufe of a fufficient quantity; but this he did not do; his quantity I think was no more than fifteen bufhels to an acre, which he fowed over the fprouting barley, juft as its green blades appeared above ground; whereas, had he fown ten bufhels on every half acre ridge land, he had compleatly manured all the barley ground. However, the foot he made ufe of did him a great deal of fervice.

This is alfo a moft excellent dreffing if that quantity is fowed on barley juft harrowed into the ground, and will not only return a good crop of this, but alfo another of peas, beans, thetches or lentils, the fucceeding feafon: on new fown turnips there is nothing better than foot, becaufe it deftroys and keeps off the infects, and vaftly forwards the turnips growth.

SECT. III.

Soot on Grafs.

IT was on the second day of *February* that a gentleman had one hundred and sixty bushels of soot laid on his meadow-ground, which killed the moss, and increased the crop of grass for four years after: but it has a little fault belonging to it, and that is, the soot is apt to imbitter the hay that is made of it the following summer; but is excellent for saving and bringing forward late sown wheat: and also, if soot or coal-ashes are sown over a second year's crop of clover, it may bring on a full bite in *April*, and cause you to stand a chance of having double the quantity of grass, than you would otherwise have, if you did not sow any of this, or other manure, or dung; and therefore is of great value for bringing on an early cover or head of grass, that may prevent the sun's burning up the tender blades and roots of it.

Now we that live in *Hertfordshire*, at twenty or thirty miles distance from *London*, think it worth our while, in some years, to buy *London* coal-soot, even at one shilling a bushel charge, when at home, to lay on our meadow-ground; for, on this, the soot will not be washed too deep into the earth, as it sometimes is, in loose ploughed land,

Chap. XII. OF SOOT.

land, but will help to preferve the grafs in a growing condition, againft frofts and chills, kill worms, grubs, dars, and flugs, or make them quit the upper part of the grafs ground, and fo prevent their ufual damage; will burn up and deftroy mofs, the common greateft enemy to meadows; will bring on a fpeedy head of grafs; and, if a wet, warm feafon follows it, in due time there will be a vaft burthen of the beft of grafs. Nothing is known with us to come up to this manure of coal-foot, which as far exceeds wood-foot, as a fhilling does fixpence, and will nourifh fuch grafs-ground fo well for three years together that no other affiftance need to be given it in that time. I write from experience, as having tried the force or ftrength of moft forts of manures in meadow-fields, and therefore recommend this, as the capital, efficacious, and moft profitable of all other manures. Twenty fingle bufhels of this foot, fow one acre, but twenty-five will do it better.

SECT. IV.

Soot on Sainfoine, &c.

ON fainfoine, twenty bufhels of this fown at *Candlemas* on an acre, once in three years, exceeds all other dreffings whatfoever. Thus likewife will foot do vaft fervice to clover, fainfoine, or any other artificial graffes; fifteen or twenty bushels dress one acre for two or three years: and, if laid on in *February*, it will force on so quick a growth, as to secure their roots against the damage of summer droughts, and produce, very likely, three loads of clover, or sainfoine hay, from off one acre, besides a great return of after-grass.

CHAP. XIII.
OF MALT DUST.

IF we dress our barley-crop with malt-dust, we commonly sow but twenty-five double bushels, or ten sacks of it, over each acre thus: out of a seed-cot, a man sows it broad-cast, as soon as the barley-seed is sown, and harrows in both together; and, if showers fall in due season, there is no fear of a good crop of barley.

CHAP. XIV.
OF SALT.

COMMON salt is of a far more potent nature than ashes, lime, or soot, and like them, will cause all crops of grass and corn to grow all the winter, when others that want this help are cut off, chilled, and killed by the power of frosts, wets and winds. How greatly then
does

Chap. XIV. OF SALT.

does this noble manure help the farmer to crops of corn and grafs beyond all others, as to the quantity, quality and cheapnefs of it, when it was at two fhillings per bufhel, as it happened to be this laft year, during the exemption of the duty; becaufe it anfwered the great ends of chilling or wounding the infect, and preventing the fatal effects of colds. One gentleman, that lives in *Wiltfhire*, fowed only fix bufhels of a common falt over one acre of corn-ground, and had the beft crop in return that he was ever mafter of.

Here I fhall endeavour by the neareft eftimate, that I and other farmers can make, to fhew the confumption that is made of falt in one year by my family, and fuch a one that confifts of a wife, fix children, a ploughman, plough-boy, tafker, and maid fervant: viz.

	B.	P.	Q.
Salt for three bacon hogs, each weighing forty ftone, that take up half a bufhel each,	1	2	0
Salt for three pickle-porkers weighing twenty ftone each, that take up a peck and a pottle a-piece,	0	3	3
Salt for our beef, a quart each week for three ftone,	1	2	2
Salt for our table and kitchen ufe at a quart per week,	1	2	2
Carry over —	5	2	3

	B.	P.	Q.
Brought over,	5	2	3
Salt for the butter of six milch cows that are constantly kept throughout the year, allowing twenty four pound per week, that is sold as fresh butter at the market, a quart a week,	1	2	2
Salt for brining twenty acres of wheat,	2	0	0
Total —	9	1	1

CHAP. XV.
OF OIL CAKE.

ONE thousand of these cakes make ten quarters of meal, which will manure four acres of land well, for twenty shillings an acre; for this number of cakes costs four pounds at the oil-mills near *Cambridge*; and so great a stress is laid on this dressing, that at *Sanden* in *Essex*, upon the borders of that county, lives a farmer, who erected a mill on purpose to grind these cakes, whose powder he used instead of lime, for wheat, turnips, barley, and other vegetables, and will be serviceable in a great degree to

---

* This is not a matter of mere husbandry; but I reserve it as the particular is a curious one.
      EDITOR,

Chap. XVI. OF HORN SHAVINGS.

to the next year's crop of *Lent* grain, either ploughed in with wheat-feed, or fown on the top of that, or barley, peafe, clover, rapes, &c. It is a very fertile manure for rather more than two years. Some allow that this meal or powder will not wafh away fo foon as powdered lime, malt-duft, and fuch like.

CHAP. XVI.
OF HORN SHAVINGS.

THIS manure is taken notice of by *Houghton*, *Worlige*, and many other authors, who all write in great praife of them. One fays, that, in the year 1694, horn fhavings were then fold, in *London*, for eight fhillings and fixpence a quarter fack, and that five fuch facks, ftrewed and fcattered in furrows, before the plough, at *Michaelmas*, will very much improve two acres of land fown with wheat-feed; but do little or no fervice to hot ground. Accordingly, he fays, when he was a boy, and now was above fixty years of age, he knew a wet cold place, of fix or feven acres of ground, that had fixteen facks of thefe fhavings ftrewed on it, and it brought corn plentifully for four years together; and that then, being laid down with grafs-feed, it brought forth very good grafs, and

continues

continues so still; that the number of horners were then twenty-four, who made each of them above twenty sacks a year.

But this account is not sufficient, to instruct an ignorant farmer in the right use of this beneficial dressing, or durable manure; and, therefore, I shall endeavour to carry on this subject a little farther, and shew a larger and more particular use of horn shavings. Of these there are two sorts sold by the horners in *London*, or those who, out of ox's, cow's, and bull's horns, make combs, inkhorns, and a hundred other sorts of utensils and knick-knacks, most of whom live in *Petticoat-lane*, near *Whitechapel*, *London*.

The small sort of horn-shavings are those commonly made by the inkhorn-makers, who sell them for about eleven shillings a quarter-sack, and two sacks sow an acre. The large sort are eight shillings a quarter-sack, and two and a half of them sow an acre; but the small sort are reckoned the cheapest, because they go a great deal the farthest, by lying closer and covering the more ground.

Both these are, by some, thought not to be so proper a manure for wet as for dry land, because the larger sort, especially, is apt to hold water, and chill the roots of wheat in such wet ground, to its damage; but where the earth

lies

Chap. XVI. OF HORN SHAVINGS.

lies dry, either in vale or chiltern lands, they are a moſt excellent manure, and will do great ſervice, for ſeveral years. In our chiltern, gravelly and other dry ſoils, we uſe them in this manner: if wheat-ſeed is to be ſown in four thorough-ſtitches, or what may be called two-bout lands, after all the ground is harrowed plain, horn or glovers ſhavings are to be ſown over all the land ; and then wheat-ſeed muſt be ploughed and ſowed in with them. A clover-lay was ploughed up in the chiltern country after *Michaelmas*, into broad-lands, and let to lie all the winter, to rot the graſſy ſurface that was now turned down, and alſo the roots of the clover againſt the ſpring ſeaſon, in order to ſow the field with hog-peaſe: accordingly in *February* the farmer harrowed the ground all over till it lay plain and ſmooth; then at the beginning of *March* he ploughed the cloſe acroſs the laſt ploughing, and at the ſame time ploughed in a ſufficient parcel of horn-ſhavings, which he bought in *London*, thus: before he began to plough the land this ſecond and laſt time, he ſowed it firſt over all with the ſhavings, and then a man followed the plough, and ſowed or ſprained in the pea-ſeed, which was a mixture of the horn-grey and maple hog-peaſe; and when all was done, he harrowed the ground as plain as he could, and then the work was finiſhed. By which management

ment the seed and the manure lay together, and thus produced such a great crop of peas, that the tasker commonly thrashed a quarter, or eight bushels of them in a day. And it is no wonder, that these shavings are so efficacious in producing such a large crop of peas, for they are accounted at least three years dressing for any land they are ploughed into, because they are of a very tough, hot, and oily nature, and therefore exquisitely well suit all dry, husky lands, especially sharp gravels, as well as cold, wet, clayey earths; and as they hollow the ground very much, they are the most agreeable manure that is for a pea-crop. But this profitable piece of husbandry of manuring ground for pea-crops with horn-shavings, extends itself in particular to the fertilizing the next crop of grain that is to succeed the peas; because when the shavings lie in the ground the first summer, for assisting them in a fruitful growth, they are rotting and mixing with the earth, and so meliorate it, that one or two ploughings will prove sufficient after the pea-crop is got off, to sow wheat-seed in the same land.

Horn-shavings are a warm, spirituous hollow dressing, and are of two sorts, the finer and the coarser. The first is generally sold at twelve shillings the twelve bushel sack, and the latter for less, and are sown on wheat, rye, or barley ground

Chap. XVI. OF HORN SHAVINGS.

ground juft before the laft ploughing, to the quantity of twenty-four bufhels on an acre; that will laft fix years in the earth.

But here an objection may arife, that fuch hot, dry, gravelly ground is not fo proper to be dreffed with horn-fhavings, as the colder and ftiffer foils are. It is true, that they are moft natural for the latter fort of land, becaufe, by their hot and dry nature, they will warm fuch cold ftiff ground, keep it hollow, and help to preferve it from the damage of chills by too much wet weather, which fuch foils are very fubject to. Yet as a hungry gravel is a fort of land that often fpoils great quantities of the new fown feed of corn, when great rains fall prefently after fowing, by binding and clofing the furface fo, that many of the tender blades of wheat or barley cannot get through it; it is a proper manure to prevent this misfortune, and at the fame time, by its tough and fpungy parts, it will continue and remain a feeding dreffing to fuch a lean voracious foil, much longer than either ftubble or any other dung; and alfo, by receiving and retaining of waters, the fhavings will keep fuch land in a moift condition in the drieft hotteft feafons; a benefit of no fmall importance to barley crops. If wheat feed is to be fown in gravelly foils broad-caft, or on any dry loams, then the fhavings are to be fown thus. When

all

all the ground is harrowed plain, and ready for the last ploughing, sow your shavings on the top of it, and plough them in as shallow as possible; then sow your wheat seed broad-cast, and harrow it in, and the harrow will tear and raise up the shavings towards the surface, and mix them with the seed.

CHAP. XVII.

OF WOOLLEN RAGS.

IN some chalky, gravelly, and sand loams, it is practised to sow their wheat in broad-lands, and woollen rags, at the same time, chopt small, and plough both in together; and thus one ploughing performs both, which is good husbandry, because such chopt rags lying in the same stratum of earth with the wheat roots, supply, in some degree, a watering pot; for, on every shower of rain that wets these rags, they receive and retain such a supply of moisture as nourishes the wheat-root a considerable time; so that the dryest summer cannot burn or dry the wheat-roots so much as to spoil their crop. And this it the more surely performs, by reason the cover of earth that lies on the roots contributes to shade them, and shelter the rags from the vehemency of droughts.

Chap. XVII. OF WOOLLEN RAGS.

In order to fow wheat in a chalky foil, after the laft ploughing but one is performed, all the ground is to be harrowed plain, and then, if the land is to be dreffed with woollen rags that have been chopped, they lay three or four hundred weight of fuch rags on the fame.

Some will drefs and meliorate chalk and the chalky loams with rags chopped fmall in the firft place, and then fown out of a feed-cot all over the ground, about five hundred weight on an acre, the beginning of *June*, and then ploughed in broad-lands, which will warm and enrich the earth againft the wheat feafon, and afterwards receive and hold the rains fo as to keep the roots of the corn moift in this dry foil, and by the heat of their woolly fubftance, help it very much againft the rigour of frofts and cold winds.

When your chalk, chalky loam, fandy loam, or other dry foil, fit for this purpofe, is harrowed plain, ready for fowing and ploughing in wheat-feed, then empty eight facks of chopped rags on one broad-land acre, each fack containing fifty-fix pounds weight, and each fackful to lie in one heap, at fome diftance from another. When this is done, let a man fill a feed-cot with them, and fow the rags broad-caft, with his hand, over the ground, and fo on till the furface of the acre is covered; then let a man directly fow his wheat-feed, out of a feed-cot, all over the fame land,

and plough both rags and wheat-feed in together.

A chalk, a gravel, and a dry loam, or a mixture of all, had the ufual yearly dreffing beftowed on fome parts of it, of woollen rags chopt fmall, fown, and ploughed in, both for wheat and barley crops, as the beft fort of all other dreffings; becaufe rags always agree with this dry land to admiration; for, if but one great fhower of rain fall on them, they will lodge and retain a confiderable deal of its water.

If we drefs our barley crop with rags, it is commonly done when it grows on chalky, gravelly, or fandy loams; here the woollen greafy rags will do vaft fervice, if they are firft chopped fmall, and fown over all the land broad-caft out of a feed-cot and a man's hand, and then ploughed fhallow into the earth; which as foon as done the barley-feed is to be harrowed in; and, as they are of a hollow fpungy nature, the rags will lodge the rains and dews, and become a fort of watering-pot to the roots of the barley and the earth about them, and adminifter fuch a moifture to them in thefe hot, dry, parching foils, that the barley-crop will grow with great vigour and expedition in the hotteft fummers, and not only be ferviceable to the firft crop of barley, but will inrich the ground for two or three years; at leaft the rags will become a dreffing for the next year's crop of peafe, oats, thetches, or tills;

for

for woollen manure agrees fo extremely well with this fort of dry land, that the *Ivinghoe* and *Edgborough* farmers buy great quantities of it every year, fome from *London*, others from the north, out of *Bedford* and *Northamptonſhires*, for thus improving wheat and barley crops that grow in their pooreſt chalky grounds, even where an acre of fuch land is let for half a crown a year; and unleſs the farmers are at this extraordinary charge for a manure, befides their fold-dreſſing, it would hardly, in fome years, be worth ploughing and fowing.

CHAP. XVIII.
OF CRAG.

ABOUT *Woodbridge* they make uſe of a ſhelly marle, which they there call Crag, which about thirty years ago was found out by mending a cart-way with it, where, afterwards happening to fow grain in the fame place, it proved a better crop than ordinary, and ever ſince they dig it for a manure. It is a reddiſh ſhelly earth, which being laid on to the quantity of twenty-five cart-loads on one acre, dreſſes it for feven or twelve years; fo that, at this time, they have dug pits of it in many places, and carry it to great diſtances, where it returns them prodigious burthens of grain in their hungry sandy grounds. In this earth cockle and other shells are commonly found mixed, which has caused some to imagine that this ground was formerly gained from the sea; and the rather, because the salt water is at this time not far from it; as it is said of the most fertile *Romney-marsh*, whose bottom seemed to me to be of a shelly earth, which has been gained from the sea not many years since, and produces such grass which exceeds all others in *Kent*, for fattening cattle in a little time without rotting them.

CHAP. XIX.
OF SEA-WEED.

IN *Cornwal*, in the winter and spring seasons, when storms mostly cause the waves to tear up the ore-weed from the bottom of the sea, and cast it to the next shore, they clamp it to rot alone, or mix it with sand, mould, or lime, which will in time so attenuate and divide this gluey substance, as to reduce it into a much shorter nature, and fit it for ploughing in, in *July*, or afterwards, for a wheat or barley crop, &c. as being one of the best of dressings for all sorts of arable lands; and so are their pilchards that sometimes throw themselves on their shore in prodigious quantities.

CHAP. XX.
OF HAIR.

WHEN cows, or hogs-hair, is bought of the tanner or butcher, it is commonly sold for eight-pence a bushel; and twenty bushels scattered over an acre of land, will be sufficient for any crop that is to grow in the same.

CHAP. XXI.
OF PIGEONS-DUNG.

PIGEONS-DUNG, is indisputably the hottest dung for all wet clays and moist loams, where, in a peculiar manner, the fiery salts act most potently in bringing forward the growth of all grain, grass, and trees, that are sowed and planted in these chilly cold grounds, and is therefore sent many miles by the vale-men, and bought at ten-pence the heaped bushel, chiefly for their barley crops, because in wheat-grounds it does not so well answer, if mixed with the earth about *Michaelmas*, for then by its lying in or on it near twelve months, the rains are apt to wash away its light fertile properties too soon; but if sowed the latter end of *February*, or the beginning of *March*, this dung will prove most efficacious if ploughed or harrowed in with the grain,

grain, or sowed out of the seed-cot on the surface immediately after the barley is got into the ground; however, in my opinion, harrowing in is the best way, for then it lies the nearest to the kernel's body; and if a wet time should succeed, the grain will soon get a head that will cover its roots, and shade them against the scorching heats of summer: but though I mention only clays and loams, yet is there no one soil as I know of, that this dung will not do good on, both in arable and meadow grounds.

Hens-dung, and all manner of other fowls dung, are of the very same sort of service to all land, though not quite so hot and good as the pigeons, if they are throwed on about *Christmas* or *Candlemas*, for then it is to be hoped sufficient quantities of rain will fall before the spring is over, as to cause all corn and grass under it to get a forward head and shelter against the summer droughts. This hen-dung, with us, is sold for six-pence the double bushel, put into the measure as hollow as can be, but if trod in and heaped, then a single bushel is reckoned at the same price, and is used in all the intents and purposes as pigeons-dung is. Hen-dung is frequently mixed with chaff, malt-dust, short horse-dung, cows-dung, &c. by putting any of them under where these fowls roost, that they may be incorporated and rot together, for here this manure

Chap. XXI. OF PIGEONS-DUNG. 103

nure will mix with, and diffuse its salts into their several species, so as greatly to encrease a large quantity of excellent hand-dressing for all sorts of corn, grass, or wood-land. This makes some reckon one load of neat hens-dung to be worth six of some common dung, and is of such light clean portage in sacks on horse-back, that it may be carried and sowed on distant grounds where the cart cannot conveniently come.

When they make use of pigeons-dung, they sow it by the hand out of a seed-cot, and, as soon as they have harrowed a ridge-land, they sow about ten bushels over the half acre, and then immediately sow their wheat-seed, and plough all in together.

Pigeon's dung is certainly very much coveted by all husbandmen, but all cannot have it, tho' they willingly give ten or twelve-pence a bushel for it, and fetch it ten miles an end.

I have this useful remark to make known, that pigeons-dung made use of from a dove-loft, is esteemed to be half as good again as that made from a dove-court, because, in a court, the dung lies so near the ground, that it draws much of its quintessence into it; which in a loft it is delivered from, as falling and lying on a boarded floor; which item may serve as a valuable consideration to all those farmers who buy and fetch this pigeons-dung

H 4

at some distance from their houses; a thing commonly done to help out with that made at home. On this account I have known an *Aylesbury* vale farmer drive his waggon above ten miles an end, and bring home fifty bushels at a time, at one shilling each, of this dung; and, as soon as the barley-seed is sown, a man with a seed-cot full of it, strews or sows it out of his hand, to the quantity of twenty bushels over each acre of land, and then harrows it in with the barley-seed. This manure, for many years past, has been esteemed the richest of all others for fertilizing all crops of corn it is sown amongst.

CHAP. XXII.

OF RABBITS-DUNG.

RABBITS-DUNG is of so hot and fertile a nature to both ploughed and sward ground, that it is sold with us for six-pence the single bushel trodden in, after it has lain near twelve months under cover, rotting, and is most commonly made use of in our gravels, chalks, and loams, where it is first sowed by the hand out of the seed-cot, and harrowed in with barley, but will not so well answer in wettish soils, as the pigeons and some other dungs will, because this is not of so hot a nature as they are; yet as this

creature,

creature, like the sheep, can live without water, its dung and stale is more hot and saline than many other of the four-footed sort are; that renders it of exquisite service both in arable, and grass grounds, if laid on either of them early enough, as at *Christmas* or *Candlemas*, that the sun and air may not dry and exhaust it, before its fertile salts are communicated to the roots of the corn, grass, or trees; likewise for the encrease of turnips this is of vast service, if thirty bushels are harrowed into the ground with the seed. It has often moved my regret, to think of the great quantities of this, as well as several other rich dressings, that are carried to the *London* laystalls, and there consumed to little purpose; for it is certainly worth our while to pay very well for this tame rabbits-dung there, and bring it thirty miles into the country afterwards.

CHAP. XXIII.

OF DOGS-DUNG.

A Farmer enjoyed great crops by inclosing a piece of ground with moveable hurdles, and dressing the same with dogs-dung, which is accounted the next best dressing to pigeons-dung; he breaks setting-dogs, and is a game-keeper, he keeps such a number of these animals, as return him

him a confiderable quantity of this dung every year, which enables him to fow horfe-beans, wheat, barley, and peafe alternately: and this year, 1741, he had one of the beft crops of wheat that fucceeded his laft year's bean crop. And what was very particular, when moft bean crops miffed throughout the nation in 1740, he had a moft plentiful one on this dry gravelly foil, and which I was an eye-witnefs of. And thus I know a gentleman, who keeps a large pack of hounds, have the greateft crops of grain wherever he fpreads their dung.

Dogs-dung is often experienced and found to be the moft fertile dreffing of all the quadruped forts, as is alfo beft known to thofe gentlemen who keep a kennel of hounds, and employ their dung in ploughed lands. At *Dagnal* in *Bedfordfhire*, there lives a publican who is game-keeper to a gentleman, which obliges him to keep a number of fetting-dogs and fpaniels, that return him a confiderable quantity of dung, in a year; and, as this man is owner of a fingle acre of land near his houfe, he dreffes it with his dogs-dung every year, which helps him to fuch crops of wheat, barley, beans, and peafe, as excel thofe of the neighbouring farmers, and thefe on a gravelly foil. But, what is moft to be admired, this man gets, by the richnefs of his dogs-dung, crops of corn every year, without the interval of

a fallow

a fallow one, though he fows his feed in the promifcuous, old, *Virgilian* way; and, in the dry and exceffive cold fpring of 1740, when moft crops of beans fuffered by it, this man's crop looked frefh, and went on fo well, that at harveft, he had the beft of crops, as I and many more beheld with admiration.

CHAP. XXIV.

OF HUMAN ORDURE.

FOR want of knowing how to manage this hot dreffing, it has difcouraged many from ufing it, on this account therefore, I fhall here make known a gentleman's method in the improvement of it. He had his houfe-of-office emptied, and the foil carried into a hovel-place, to remain under cover in a heap, which as foon as done, he had it all covered with mould, where it lay fome time, till the dung got dry. Then he got more mould, and chalk, or lime, and had it mixed all together with the fhovel, and let it lie fome time, when he had it turned and mixed again, and, by this, he got it into a dry and almoft powdered body, fit to fow out of a feed-cot; and accordingly he had it fowed over his wheat in the month of *January*, by a man's hand as we do foot out of a feed-cot, and it produced

wonderful

wonderful crops. When cured after this example, it is one of the best of dressings, both for corn, as well as grass-ground.

CHAP. XXV.

OF LIME.

THE salts of limestone were fixed in the crude stone or chalk, and unable to act in the growth of vegetables; but being made free, and set at liberty by calcination, they act; for till the acid barren quality of stones or other minerals is evaporated by fire, the salts are of little or no signification to land; and so of chalks and other earths, unless they are cured by fire, fermentation or putrefaction, they are not capable of benefiting the ground, grain or grass they are laid amongst, because no hard dense bodies can mix with, and assist the earth, as the more loose and powdered ones do, nor can their worse sterile earthy parts, that lie concealed in their original pristine state, be of service, till they are converted into a contrary nature, which makes this dry burning lime not only absorb and consume the wets, and moisture of grounds that are often fatal in the winter and spring to the wheat, rye, and grass, but also kills, wounds and drives

away

Chap. XXV. OF LIME.

away worms, snails, grubs, and caterpillars; lime therefore is used many ways in farming. First, by mixing it with turf, dung or mould, and suffered to lie till they are rotted and incorporated together, which will make an excellent dressing for corn or grass-ground. Secondly, forty bushels of stone-lime should be slaked in one heap in the field by the weather, which it will do in a few days, and then sowed over an acre out of a seed-cot as thin as it can well be; this may thus remain for a week, when the wheat may be harrowed in broad lands, or sowed in stitches: others will fetch water directly and slake the lime as soon as it is in the field, and sow it while it is hot, that it may bring the ground into a ferment. Thirdly, it must be in like manner and quantity sowed over the ground with turnip seed; the lime is first to be harrowed in twice in a place, and then the turnip seed once in a place with light harrows. Fourthly, On a clover lay, forty bushels may be harrowed in while it is hot, and let lain a week or fortnight, then harrow in the wheat; this will deliver the grain from the destructive worm and other insects, that often here destroy the wheat, and will also cause it to gather thicker, and maintain itself in the cold wet times of the winter and spring, when others sicken or die; and therefore I am surprized to find a late author say, that

lime

lime is most natural for light, sandy, hot, gravelly grounds, and bad for wet, cold gravels, but worst of all for cold clays.

The practice of manuring, dressing, and ploughing many of their open common field lands, lying here, I am going to shew, because, though their ground is situated as low as those in *Aylesbury Vale*, yet there is a considerable difference in their management. Many of these farmers may be called good husbandmen, and it concerns them to be so, as they pay seven shillings and sixpence an acre for their arable ground, that lies in ridges and furrows, and like those in *Aylesbury Vale* is never ploughed a-cross; but then they enjoy some of the best of ground, for there is hardly any exceeds it for producing large crops of wheat, barley, and peas, &c. as being for the most part red, clayey and gravelly loams, which lie in rood and half-acre lands, that they dress and manure with dung, with the fold, and with lime, &c. for a wheat crop, with rotten dung; and then they plough it in at the first stirree-time, and seldom or never just before the last ploughing, because here their soils are very apt to breed smutty wheat; others mix highway dirt or other rubbish with their dung, and, after it is well prepared, lay it on their wheat land. A certain gentleman, in this country, was so curious a

husband-

Chap. XXV. OF LIME.

husbandman, as to collect ant-hills, tanners-bark after its being made use of and done with by the tanner, and privy-dung, and with his stable dung made a compost, which he would let lie in a great heap, or dunghill, some time, till he had it turned; and this he did twice or thrice till he had kept it a year or more, and made it all so fine, that it would pass through the large wired screen; when it was thus got ready, he applied some of it to his garden use, and some to the field use; if to the latter, it was carried in a cart, and, when on the spot of ground, they threw it out, by spreading it over the land with a shovel, and then sowed and ploughed in the grain. As to the fold, there are hardly any but great farmers that employ one, and that a little one, because this being a flattish, wettish, open field country, the sheep here are very subject to the rot; and, therefore, if a farmer has not the conveniency of shifting them in meadow ground, or on ploughed lands that lie dry (which few small farmers have) the loss may be more than the profit. There were very few sheep killed by the butcher in the dry year of 1742, but what had been formerly touched with the rot, as appeared by their hard cory liver; else this field-dressing would be of excellent use here, as agreeing very well with their sort of soil. But their lime

lime is used in general, and answers exceeding well on their gravelly loams and gravelly clays; contrary to those ridge, loamy, clayey lands in *Aylesbury Vale*, which being of a stiffer nature, and lying higher, do not suit this sort of dressing, nor did I ever see any make use of lime in this vale; although a great deal may be said on account of prejudice, which not a little governs the countryman, as is obvious by the non-use of lime, in the large open fields of *Pitstone*, near *Ivingboe*, where they have hundreds of acres of gravelly, loamy land, that would rightly agree with the lime improvement, and yet pay no regard to the great opportunity they may enjoy of making lime at a cheap rate, because under all their surface, they have a chalky bottom: but, instead of lime, they go sometimes ten or more miles, to buy and fetch rags home at a much dearer rate. But this vale-farmer is wiser for liming his land; and I must here observe, that in *July*, or *August*, they take care to lay on the just quantity of two quarters and a half upon each half acre ridge-land, in little heaps, over the whole land, to lie and flake, till all is ready to be spread, and then to be ploughed in at the second stirree-time. Others lay a whole waggon-load of lime, which is five quarters, in one heap, on a land, to lie a month or more, to flake and crumble by the weather,

and

and, when ready, they carry it in a little low cart, and by a shovel take it out and spread it over two half acre lands, which after having been ploughed in twice or more, they sow and plough in their wheat-seed, and then they ridge up their land, contrary to the *Vale of Aylesbury* practice, for here they always cast down their land at sowing time. Some of these farmers will also lime and dung their land the same year, which, being a double dressing, generally returns them prodigious crops of wheat and *Lent* grain.

CHAP. XXVI.
OF HOOFS.

HOOFS, of oxen or cows especially, instead of cutting them to pieces and ploughing them into the ground, we force them into the earth with sticks, presently after the wheat, barley or rye is sown, at about six inches or a foot distance, in order to prevent dogs, &c. carrying them away; here they lie almost smothered, ready to receive and lodge the rains, which will cause them to rot, putrify and mellow the ground; so that by reason of the dirt about them, their fast lying in the ground, and the nauseous taste and scent, they lie safe from dog, fox, badger, wild and pole-cats, weasels, ranes and stoats, and is a dressing for six years, to chalks, gravels, and loams, where they will prove so many little watering-pots, or receptacles, to such dry grounds, and exceedingly nourish them.

CHAP. XXVII.

OF HOGS HAIR.

HOGS hair and coney clippings are very good dressings for the light soils harrowed in with wheat, rye or barley; so are oil cakes of rape-seed, &c. that are now used much about *Luton* in *Bedfordshire*, which they chop small or grind, and plough or harrow in, to a great profit. These are most of them general manures, and used more than ever, insomuch that I know four farmers, neighbours, that have equally contributed to the purchase of twenty twelve-bushel sacks.

CHAP. XXVIII.

OF BUCK WHEAT.

IT is the opinion of some, that it does most good in clays and clung grounds to hollow them; others say, that it is best used in dry, binding gravels

Chap. XXVIII. OF BUCK WHEAT.

gravels or sands, to keep the one loose, and the other moist; but in some of the sandy lands about *Godalmin*, they dare not sow it, lest it make the earth too light for common wheat; yet in heavy sands, sandy loams, gravels, and such sort of light grounds, it is certainly best sown, either to plough as a dressing, or for seed: In chalks it is improper, because it loosens them too much. In stiff and wet lands, it is apt to rot both in root and ear, as being itself a very succulent plant, and of a tender, cold nature, being easily hurt by some winds, lightning, and frosts; and therefore warm light soils suit it best, and, in all, it is a very great killer of weeds, where it grows thick, as it commonly does in tilths, though you sow but one bushel on an acre, and is found by many to answer that great end, which all farmers endeavour after; namely, to reap a great advantage for a little charge. I am the first that have sown it in our parts for a dressing, though it has been sown in *Norfolk*, *Suffolk*, *Surry*, *Kent*, and *Berkshire*, many years, and is now more and more propagated. As I live twenty-seven miles from *London*, it is a great expence to have manure from thence; for sometimes soot is one shilling a bushel, so that twenty-five shillings is the charge of dressing a single acre well, when two shillings and sixpence will do in *French* wheat, and yet
this

this will kill weeds when foot will nourish some; the wheat dressing will last three years*, when the soot will last but two; the one is also more certain than the other, for in dry seasons, the soot by its sulphureous quality will burn up a crop of wheat or barley to a great degree, when the *French* wheat will remain a sure and certain promoter of vegetation, by its moist, hollow, fertilising nature: again, soot, when it is laid to the quantity of twenty-five bushels on an acre, forces so vehemently either grass or corn, that, after two years, it leaves a poverty behind it, which this is so far from doing, that it is almost a dressing the third year, provided it be a full crop at first, and ploughed in at full bloom, when the stalks are fullest of sap.

This grain has returned four or five quarters of seed from one acre, which is excellent for feeding hogs and poultry. In blooming time, when this is to be ploughed in, it is very serviceable for subsisting cattle in a little while, who will tread it down, and make it the fitter for this purpose. You, that dress your land by sheep, may fold them in another ground, while this is provided for, and that without any risque of your cattle's breaking out, into your neighbour's

---

* This assertion of Mr. Ellis's, is to be doubted.

fields, for here is meat in plenty. So, when your teams are all busy, in some ground there needs but one ploughing, to harrow your common wheat in, on this *French* wheat.

This good piece of husbandry may be performed to make it answer very profitable ends on several accounts: one whereof take as follows, viz. In *Norfolk* I saw this done, the latter end of *August*, where, when a crop of *French* wheat was just in bloom, they with one foot-plough ploughed it in; and, with another, that directly followed, they turned up a sandy mould on the first layer, and made the whole field appear in one even surface of such fine earth. On this they forthwith sowed two bushels of rye on every acre, and harrowed it in. The same might be done with common natural wheat, if the soil is proper for it, and many plough in *French* or buck-wheat for the same purpose of nourishing a common wheat-crop.

CHAP. XXIX.
OF OLD THATCH.

THIS sort of dressing is now found to be of good service, when put over sward, or natural grass land. In *Billingter* large common field meadow, lying near *Leighton* in *Bedfordshire*, he that has saved his old thatch, will be sure to lay it on that part of the ground belonging to him; as soon as his grass is mown off, and expect it to do him great service, by shading the roots, and preventing cows, horses, and sheep, biting down the after-meath, or grass too near the roots, which, when the latter sort in particular do, it commonly affects the next year's crop, to the great lessening of it: whereas when they lay on a good quantity of old thatch, they seldom fail of a considerable crop the succeeding season, for, by this means, they have mowed nine loads off four acres, at one mowing-time, in this open common field. To this account, I add, that thatch very much hollows the ground by its cover, and by the worms pulling it in; and thus it lets the dews and rains have the freer access to the roots of the grass, to its great improvement.

CHAP. XXX.

OF THE COMPARATIVE QUANTITIES.

BESIDES the dung and urine that the cattle leave in the turnip-ground, and after the seed is sown, short rotten dung should be immediately spread all over it to the quantity of fifteen

cart

cart loads on each acre, at leaſt; or twenty buſhels of ſoot ſown over the ſame by the hand broad-caſt, or three or better four cart loads of clay, or wood-aſhes, that will contain in all about one hundred and twenty buſhels, reckoning thirty buſhels to each cart load, to be ſown out of a feed-cot broad-caſt over one acre of ſuch new ſown wheat ſeed: or ſeventy buſhels of coal-aſhes, or twelve buſhels of peat-aſhes, if they are made from the beſt, black ſtrong moor-peat; or twenty (or rather leſs) buſhels of oil-cake powder, or forty or fifty buſhels of malt-duſt; one of which powdered manures is to be ſown by the hand out of the ſeed cot, as ſoon as poſſible after the wheat-ſeed is ſown, for the rains to waſh in upon the corn; to force on ſuch a ſpeedy growth, ſow forty or ſixty buſhels of ſlaked lime over the acre of ſeed, or twenty buſhels of ſoot, or ſixty buſhels of coal or wood aſhes.

As ſoon as young coleworts appear above ground, ſow over every acre twelve buſhels of peat aſhes, or forty of coal aſhes, or ſixty of wood aſhes, or twenty of ſoot,

CHAP. XXXI.
OF MANURING MEADOWS.

IN *May*, be increaſing your dunghills, in order to get them rotten and fine againſt the time you get your field cleared of the firſt and ſecond crops of graſs; ſome lay it on after the firſt to increaſe the ſecond; others not till *October*, when it is fed bare: this ſort of dunghill requires to be turned often, mixed well, and rotted to the higheſt perfection of fineneſs; for the ſmaller its parts are, the better and ſooner it mixes with the ſmall roots of the graſs, and thereby brings on an early cover, and that an early mowing, once if not twice in a ſummer, and then great burthens, as is annually proved by the *Edgware* and *Hendon* men in particular, who certainly are the moſt curious managers of a dunghill in this nation, and yet are at a conſiderable charge all the ſummer, to bring their *London* lay-ſtall, muck, and dung, and coal-aſhes, three, ſeven, and ten miles on the wheels, which they moſt carefully mix with highway dirt, mould, and ſometimes chalk, and incorporate them ſo often till they attain a moſt exquiſite fineneſs againſt their ſpreading ſeaſons.

CHAP. XXXII.
OF THE VARIATION OF MANURES.

THIS is abſolutely neceſſary in both meadow and arable grounds; for in the firſt it it has been proved, that by sowing ashes some years together, they have lain at the roots, violently drawed the goodness out of the ground, and in time so choaked the plants, that they produced but very thin crops of hay. So horse-dung, for the same reason, may be annually applied, till the ground will bear no grass. Arable land will also complain, if too frequently dressed with one sort of manure. As for example, if soot or lime is used each on two or three crops together, the consequence will be thin returns of grain; a judicious farmer therefore always takes care to change his dressings as often as he can; hence it is that many are so curious to mix earth, lime, and dung together, as best agreeing in the dung-hill, and on the ground afterwards. Again, the nature of manures is to be considered for their alternate application. Soot, lime, and ashes generally produce a short thick grass; dung or fold, long grass; for which reasons, dung or fold should follow soot, ashes, or lime, and so on; likewise where grass on downs, or other dry bottoms, is naturally short, cart-dung or the fold should be made use of, before the burning sort. A gentleman that keeps twelve hundred sheep lays so a great stress on this piece of husbandry, that he hopes, by this means, to have sheep worth sixteen or eighteen shilling a head, instead of a smaller sort of eight or ten shillings value, which

he

he used to have; and that by employing his fold on such short grass, and feeding his sheep in the winter with pea or other straw out of racks; for either hay or straw, consumed in this manner, warms the roots of the grass, keeps off chills, prevents the growth of moss, and, by the worms drawing it into the ground, becomes a sort of under dressing, besides the improvement of the dung, stale, and heat of the sheeps woolly bodies.

But above all other management, there is an absolute necessity for making use of these dressings alternately, that is, not only to dress the grain, grass, and soil with a manure, but to alter these supplies each season; for if I dress my wheat crop this time with stable-dung, the next time it should be with soot, and the third time with the fold, horn-shavings, hogs-hair, hoofs, or some other change, and so for any other grain, &c. else it may be depended on, the earth and grain will be saturated, and tired with the repetition of one and the same sort, which is often the cause of thin crops, and to my knowledge but lightly regarded by some ignorant husbandmen; though to my satisfaction I must own this profitable notion of late has obtained a probation with many, that formerly cavilled with a useful tenet meerly for its being new; for undoubtedly different dressings impregnate the ground with different salts and sulphurs, which oblige the roots

of

of all vegetables that grow therein, to draw their food each time in a various manner.

CHAP. XXXIII.

OF THE IMPORTANCE OF MANURING.

TWO farmers have got eſtates out of a poor hungry, gravelly, chalky, earth.— They live at this time in *Ivingboe* pariſh, which is ſo large as to extend its bounds into the very low vale of *Ayleſbury*, and into the very high *Chiltern* country. It is in this laſt that two farmers live, who rent about ſeventy pounds a year apiece; one whereof is a quaker, and both live ſo cloſe to each other, that ſome of their lands join; and they go on after one way of farming to very great profit, for each of them keeps about three hundred ſheep to fold on their poor gravelly, and chalky grounds, which they are enabled to do by the help of *Ivingboe* common and their own incloſed fields. But this is not all by a great deal, for theſe knowing farmers carry their good management much farther, being of opinion that by laying out the value of a ſingle dreſſing, they get the value of a treble one. To explain this will be very eaſy, when I make known their expence in buying large quantities

of

of *London* soot every year, besides other dressings; for as their farms lie thirty miles distant from *London*, every bushel of soot costs more than three-pence carriage, which, with six-pence a bushel for the neat soot (provided it is bought in summer against the next spring) costs nine-pence in all; but by the worse manager it is bought in *January*, *February* and *March*, for eleven and twelve-pence a bushel, carriage included, though it is delivered at *Hempstead*, two and twenty miles from *London*, from whence they fetch it in their waggons or carts after they have, the same day, delivered their corn to be sold at this great market; so that notwithstanding this back carriage of their soot, perhaps eight or ten miles by the farmer's own team, and paying eight-pence a score as we call it, that is, for sowing twenty bushels of it over each acre, they think it well worth their while to buy and sow it on some of their several sorts of ground every year; which, with their folding, causes their crops of wheat and barley to rally in the cold pinching spring season, when other crops are declining instead of flourishing, that go without such a beneficial warm assistance. But this is not all; these farmers buy rags, horn-shavings, hooves, cows and hogs-hair, &c. and apply one of them to their hungry gravelly ground, which prospers most of all under the help of these tena-

cious,

Chap. XXXIII. *of* MANURING. 125

cious, durable, and tough dreſſings; for, as a gravelly ſoil is compoſed in a great meaſure of pebbles, and very ſmall ſtones, it proves a ſort of ſtrainer or drainer, to powdered or other thin dreſſings, which are ſoon diſſolved or carried down below the roots of the corn, by the repeated waſhings of great rains, unleſs they join the fold or other dreſſing to them. In ſhort, theſe two farmers have got good eſtates, meerly by their well managing a ſeventy pounds a year farm each; for, in ſuch poor ſoils, a penny kept back from being laid out in dreſſing, is two-pence or three-pence loſt. This plainly proves the abſurd, obſtinate old farmer, to be notoriouſly wrong in the management of his gravelly farm, becauſe by not dreſſing his wheat and barley crops ſufficiently, his after *Lent* crops are ſeldom worth more than twenty or thirty ſhillings an acre, and ſometimes not that; he, therefore, ſince his farm is about the bigneſs of theirs, ſhould buy every year, like them, ſix hundred buſhels of *London* ſoot, beſides other dreſſings, to be a compleat manager of ſuch a hungry farm; and then neither he nor any other who does the ſame, will ever come under that great loſs which befel a young farmer, about the year 1736, at *Buttermere* in *Wiltſhire*, ten miles diſtant from *Newbury* in *Berkſhire*, who, being juſt married, laid out

moſt

most of his wife's fortune in stocking a large chalky farm, but neglected very much the dressing part; insomuch, that after sowing forty acres with oats in one spring-season, he had such a poor crop in return, that he thought them not worth mowing; and, therefore, turned all his horses into them to feed, as on so much grass. The consequence was, that he broke, I think, the same year, as I was informed on the spot the year following, when the landlord of the same farm was obliged to take it into his own hands. This particular case, as well as the consideration of several others, leads me to praise the most excellent method of purchasing dressings from *London*.

This also declares the folly of those who hire farms, and, at the same time, are incapable to give the land its due manuring; however, whether it is want of knowledge, a willing mind, or ability, the deficiency will surely be a loss to the occupier; accordingly, I have known able persons run out considerable sums of money in farming, for want of a right notion of the value of manures and dressings. Others begin at the wrong end of farming, and allow themselves in great expences, to erect needless new buildings or adorn old ones; make large gardens, fish-ponds, visto's, &c. at the same time neglecting the improvement of their land, that ought to be

first

Chap. XXXIII. *of* MANURING.

firſt remembered to be duly aſſiſted with proper manure; that would by them produce, and become a ſort of fund to anſwer a number of contingent charges.

Soot we ſhoot down in our fields in a round pecked heap, againſt next ſpring, when we ſow it over our wheat; and which, in the mean time, by being hurdled round, and a little ſtraw laid on its top, will keep free from damage of all weathers. The trotters, hooves, and horn-ſhavings may be ploughed in at the firſt ſtirree-time in this month, and anſwer extremely well for ſeveral years after. But the rags, hair, and leather-ſhavings ſhould be ploughed in at wheat ſowing-time, as the aſhes are to be ſown out of the hand on wheat, clover, ſainfoine, and natural graſs, in *December*. Now all theſe, and more, we *Hertfordſhire* farmers make uſe of, as we are an inland county and abound with many ſorts of ſoils; and I believe I may aſſuredly vouch it, that he who rents from fifty to a hundred, or more, pounds a year, and does not make uſe of ſome of theſe manures and dreſſings for his meadow or ploughed grounds, cannot get money in any conſiderable degree, notwithſtanding the conveniency of commons for his ſheep, and other home advantages. For a proof, in a great degree, of this, there are three farms within two miles of my houſe that join together, and rented

by

by three several tenants, called *Wards*, whose soils are loams, chalks, and gravels; I say, here they often lay out the value of an ordinary crop in *London* manures, before they can obtain a good one. Indeed, when a very dry spring and summer happen, they are liable to lose money instead of getting it, by the burning quality of these dressings; but when they are attended with sufficient moisture, they commonly have double crops to what their neighbours have, who made use of none of these auxiliary helps; for though the former are subject to some fatal incidents, by weather and insects, yet it is well known that no farmers in our parts get more money, in proportion to their farms, than these do, notwithstanding their lands are of the poorer sorts.

BOOK

BOOK III.

OF TILLAGE.

CHAPTER I.

OF PLOUGHING.

IN vale ploughing, ridging up is done by beginning in the middle of a half land, or half acre (in which fashion most of their vale grounds lie, for the greater convenience of carrying off the water) with the foot plough, which, the first time of fallowing, fills in the great middle thorough that was left when the last grain was sowed thereon, by drawing the plough up and down each side of the middle of the land, till the half acre is all ploughed. In this mode all their ground is ploughed till they sow their grain, and then they alter.

Casting down is only done the last time when they sow their seed, by beginning at each outside of the half acre, and ploughing every thorough

rough down, till they come to the middle, when they leave a large thorough, wide enough for a single horse to go along without offending the corn, if it is not too high.

For wheat they begin to plough or fallow up the bean stubble in *April* for the first time, and harrow it down before they plough it again, which is all the harrowing they give their wheat land; then about *Midsummer* they plough it again, and let it lie till the middle of *September*, when they plough and sow the last time for good; but the more industrious man will give his wheat land a fallow in *April*, and two stirrees between that and sowing time, so that in all there will be four ploughings. Now the fallow ploughing is performed by ridging up, and so is all the two stirrees after the same fashion; but the last ploughing is done by casting down the land, so that they harrow but once, and that is after the first ploughing; and at last they leave a large thorough in the middle, unless it be in a very low wet piece of ground, which some always ridge up, the better to keep it dry.

For barley they plough the bean stubble four or five times in all. The first ploughing is begun in *April*, for they say, *Better an* April *sop, than a* May *clot*, which often makes the sluggard put forward to get his ground ploughed in that month; then they harrow it before they plough it

Chap. I. OF PLOUGHING.

it a second time, which is about *Midsummer*, and at *Alhollontide* they plough it a third time, and let it lie till *March*, when they plough and sow for good. Here then is one fallow and two stirrees, that are done by ridging up, and the fourth time by casting down and sowing; some of the best sort will give it a fallow, and three stirrees before they sow it.

For beans they give the wheat or barley stubble only one ploughing; by casting down the land, and harrow after sowing, which some do presently, others not till the beans appear; this is usually done at *Candlemas*, according to the rhyming proverb, *At* Candlemas *day, it is time to sow beans in the clay.*

Peas are done after the same manner, but something later.

Chiltern ploughing.

The chiltern farmer's condition varies much from him in the vale, and particularly as to his ploughs; for as the vale man uses but one sort, the chiltern-man is obliged to occupy no less than three several sorts; and they are the fallow-plough, pea-stitch, or scent-seed plough, and wheat-seed plough, besides the necessary occasion there is sometimes for the foot plough, the *Kentish* broad-board plough, the creeper, and the swing plough.

At *Carrington*, near *Dunstable*, on their high clay ground, they sow all they can in stitches to keep the corn dry, for in these the roots are more exposed to the sun and air; so in gravels and chalks, they more frequently than formerly run upon the stitch, because, in all soils, it saves the rains and dressings better than the broadland; for as the rains fall on the stitch, it washes from the upper part downward, by which the goodness gradually descends on the sides and lower parts, till it comes into the gutter between the stitches, and there meets with the bottom root: and of stitches, the lesser one is accounted best in the clays and loams, because it lies more sharp and open, for the reception of those benefits that the kind seasons afford. This way has got into such esteem at that place and *Kensworth*, that several there sow their oats in stitches, by first bouting up the land in *November*, and in *February* or *March* they sow their oats in four thorough stitches, by a man's following the plough and straining them into the thoroughs. But in gravels and chalks their stitches are something broader than in the clays and loams, as being drier soils, and are commonly made at the rate of eleven to two poles length, which, and other ploughings, I shall describe as follows, viz.

Broad-land ploughing is what is oftenest done both in vale and chiltern of any other, and is com-

Chap. I. OF PLOUGHING.

commonly the firft operation that is performed on a ftubble of corn or lay of grafs, with either the fallow or foot-plough's being drawn as clofe as can be on the ground, and turning the land thorough by thorough into a flat even fhape and form. This method being eafieft for the horfes, at the firft breaking up of the ground, commonly precedes and makes way for that of bouting, four thoroughing, and hacking, &c. and is called clean ploughing.

Bouting, or bouting up, is a half ploughing of the ground, by making a fingle ftitch either from off broad-lands or wheat ftitches; it is done as the firft fallow about *Albollontide*, to prepare wheat ftitches for fowing of barley or peas, by making one thorough, which raifes a fmall ridge of earth; then, at near a foot diftance, another in like manner; both thefe together become one ridge or bout, and is very proper for giving the froft an opportunity to kill the weeds, and fweeten the ground. Next time is the middle of *February*, when they bout it down, and harrow it plain crofs-wife, where it lies perhaps a fortnight, dunging it all over; then they plough it into broad-lands and harrow in barley. This bouting is alfo done by ploughing two thoroughs off a barley ftubble at *Albollontide*, within a foot of each other, that makes a ridge or bout to prepare the ground for peas, which is to lie till the middle of *February*, when the bout is to be harrowed down; and at the beginning of *March* they plough it the same way as it lay before, into four-thoroughs, which they sow by ſtraining in the peas, and harrow it directly down. Bouting is also performed in ſummer, by firſt ploughing the oat ſtubble into broad-lands, in the month of *April*, which is called fallowing them in *June*; that muſt be harrowed plain, and ploughed into bouts that may lie till a month after, when it is to be ploughed up again, which will clean plough that ground that was not broke the laſt time, by running the plough this time backward and forward through the middle of the ridge or bout, this will raiſe another ridge that will lie directly upon that which was a thorough before; then a fortnight or month before the ground is ſowed, it may be either back-bouted or thoroughed down, and harrowed acroſs directly, when it is in right order to ſow wheat in ſtitches.

Bouting down is done by making a ſhallow thorough on each ſide of the ridge of the bout, whereby a ſleeving, or a narrow thin ridge remains in the middle; this is done on purpoſe for the harrows to pull down, as they are drawn acroſs the land to level the ground, by preparing it for the laſt ploughing and ſowing of wheat. This is alſo called back-bouting.

Thoroughing down is a work ſometimes done
inſtead

Chap. I. OF PLOUGHING.

instead of bouting-down, or back-bouting, by drawing the wheel fallow plough only once deeply through the very ridge of a bout or stitch, and laying it in order for the harrows, that will, by drawing them acrofs, level the ground ready for ploughing and fowing of wheat; and it is performed as well as the bouting-down according as the earth requires; for if it is very fine and loofe, then thoroughing down, as it is lefs trouble, will be fufficient; but if it is not in quite good order, then back-bouting is the more neceffary, by reafon it fines the ground fomething more.

Four-thorough is performed by the wheel fallow plough, when the ground has been the time before fallowed into broad-lands, and harrowed plain; this then is inftead of a ftirree, and will fweeten the ground to an admirable degree in the fummer againft the wheat feafon, which fome harrow, and fome do not, before the next ploughing, and that is bouting up thefe ftitches, which is the third ploughing; the fourth is thoroughing down, and harrowing plain; then it is ready for fowing wheat in thefe four-thorough ftitches, with the wheat-feed wheel plough. This is not quite clean ploughing, yet is reckoned the very next to it, becaufe this way brings the earth under an expeditious finenefs and fweetnefs, by reafon it breaks and expofes moft part of it to the fun and air, by its higher fituation than the broad-land; for by this mode the land is raifed beyond its common level, and thereby can better difcharge the fugging wets that often, by their long lying on the clays and loams, four the land; this way is done by drawing the fallow plough backward and forward, till four-thoroughs are made almoft clofe together, which is a fort of ridging up; beginning firft with one of the uppermoft thoroughs, then almoft clofe to that another thorough, whofe earth will join that which came out of the firft; then on each fide of them another thorough muft be made as before, which will compleat a four-thoroughed ftitch. This is a fomething more tedious operation than the broad-land, but generally fufficiently pays for both time and trouble. The wheat ftitch lies very *apropos* for this work, and is often at *Albollontide* ploughed into four-thoroughs, that remain fo till the next fpring, when they harrow it down, and plough it again into the fame four-thorough ftitch, in which they ftrain their peas, with the wheel-pea ftitch-plough.

Hacking or combing may be called a clean ploughing; if the ground is fine when this is done, and the ploughman leaves not too great a kicker, for this way is commonly the laft operation but one, juft before they plough for good to fow wheat, barley, and turnips: it is done on broad-lands by making a thorough with the fallow

Chap. I. OF PLOUGHING.

low wheel plough, and making another so close that it throws in that earth that first came out, and fills up the thorough again; then they sometimes leave a small bit of ground, which they call a kicker or sleeving, and then make another thorough, and so proceed, by throwing that in again as they did the first. This way is a prodigious sweetener of the ground, and tears it all to pieces with the help of the harrows that directly succeed this ploughing; then the next and last ploughing is done across the land, which still adds to the fineness of the tilth.

These several different ways are of very great moment in getting ground into order for the reception of the seed, which is the main art of farming, else the weed and sourness of the grounds are apt to become master; and as the farmer is obliged to keep seasons in ploughings and sowing, he hereby has an opportunity to get ground sweet and clean by a skilful method in a confined time, which otherwise perhaps could not be done, as often is visible in grounds of turnips that are eat off by *Albollontide*, *Christmas*, or *Candlemas*. These times call for various managements; if at the first two, then the earth should be bouted up and let lain till the middle of *February*, when it should be back-bouted or thoroughed down, and harrowed level; then it is ready for the last ploughing into broad-lands, the

cross

cross way of the former, and fit for harrowing in of barley. But if at *Candlemas* the turnips are eaten off, then they often hack the land, and let it lie to sweeten till the first of *March*, when they harrow it plain, and so let it remain till the middle of that month, and then they plough it across and harrow in barley.

Others, again, at *Albollontide* or *Christmas*, will plough it into four-thorough stitches, which will much better sweeten the land and kill the weeds, than broad-land ploughing; this they let lie till *Candlemas*, when they bout it up, and at the first of *March* back-bout it down, harrow it well, and is then ready to sow a fortnight after by ploughing across.

These are certainly much the best ways if time will allow after the turnips are eaten of; for broad-land ploughing, in winter or spring especially, has been the cause and ruin of some crops, it being only a turning the ground topsey-turvey; the twitch and weeds are not so soon killed as when the land is more laid open; for in this way, as I take it, the earth is but half exposed, when bouting or hacking does it at least three parts in four, and keeps the land dryer and more to the sun and air.

But when the turnips are eaten off late, and opportunity won't admit of plurality of ploughings, then they give it but one ploughing in
broad-

Chap. I OF PLOUGHING. 139

broad-lands, and harrow in the barley or wheat, becaufe the fheeps dung by a thin ploughing, lies juft under the barley for it to root into, and have the fudden benefit thereof; this is reckoned a furer way than two broad-land ploughings, by reafon this being often a wet feafon; the danger of a double ploughing is, that if it fhould rain after the firft ploughing, it will bake the earth down, and never be fine; but if a dry fine time follows, then it is better than one, and the firft ploughing muft be deeper than the fecond to turn up again the fheeps-dung.

A farmer was at plough in a fnowy time, and having a great deal to do, would go on, and ploughed in fnow as it fell: the confequence was, that, as he was then fowing peafe, the *May*-weed came fo thick up, afterwards, that there was no peafe, only a thetch appeared here and there, that happened to be fowed among the field peafe: but this was the leffer damage, for the ground was made fo four by ploughing in the fnow, that it did not come to itfelf, that is, the farmer did not get it into a fine fweet tilth again till the feventh year. Another farmer would plough his ground while the fnow lay on it, and came off in the fame manner; for fnow, when mixed with the earth by the plough, lies a confiderable time in it before it is diffolved, and in that time it becomes a fort of cement,

cement, and serves to make the ground cling together and increase its sour quality, to the great damage of succeeding crops of grain. Besides which, in case frosty and chilly wet weather happens for some time afterwards, the very next crop of pease, oats, or barley, will be certainly a very poor one; because when the roots are chilled and starved by cold, above and below, there remains little hopes of success.

Next to ploughing ground in a snowy season, doing the same in wet weather is likewise of pernicious consequence to the farmer; for although it is not so bad altogether as the former, yet when a journey at plough is performed in a great rain, it will prove bad enough to him, because all clays, stiff loams, and gravels, will be so bound and closed, that at the next ploughing the farmer may depend on finding his ground pretty well furnished with weeds, and ploughed up in blocks as we call it, that is, so clotty and rough, that it will be little or nothing the better for the first ploughing.

I am owner of an inclosed field, whose soil is a gravelly loam, that is so apt to run into a consistence, like the batter of a pan-cake, in case a great rain falls in the time of ploughing it, or presently after, that we are forced to take special care to put off ploughing it, when there is reason to suspect a great rain would attend the operation;

Chap. I. OF PLOUGHING.

tion: but, in loose, short, chalky soils, there is not that necessity for such extraordinary care, because these do not suffer damage like the others.

A man lived at *High-street Green*, in *Hertfordshire*, and kept but two horses in all for ploughing the little land he was owner of. These being not able to draw the plough deep enough, to extirpate the roots of weeds, and to give the earth a fine loose bottom for the roots of corn to enter easily; his ground seldom returned above half crops, which so impoverished him, that he was forced to sell some of his land: accordingly he sold a field of eight acres, which after the buyer had ploughed it with a strong team of horses, and made the share of the two-wheel fallow plough enter deep into the ground, it turned up such a fresh parcel of earth, that he had forty bushels of wheat from off one acre, the very first crop, that, when the whole was sold, the sum amounted to near the value of the land. These are plain instances of the great advantages attending the true knowledge of ploughing ground in the right manner.

On the farm before mentioned, it was usual for the former tenant to give strict orders to his ploughman to plough an acre and a half at one journey, or in one day, before he came home; accordingly the ploughman did, but then he was forced

forced to plough large thoroughs, with his wide-set broad-board wheel fallow-plough, which every time turned almoſt ſixteen or eighteen inches wide of earth, and which, indeed, ſhould have been rather turned at twice, for then the ground would have been ſo broke, that weeds would have had leſs power to grow and increaſe. This miſmanagement broke the farmer, and had like to have done another near *Rickmanſworth*, but his eyes got open juſt in time; for, as ſoon as he was ſenſible of his error, he took in his fallow-plough narrower, and ſaved himſelf from that ruin which otherwiſe muſt have come upon him. For ſuch ploughing not only keeps the land four and hard, but gives weeds a foundation to breed, and grow luxuriant; becauſe, in ſour hard grounds, the roots of corn cannot ſtrike in their thready fine fibres, with that freedom and eaſe as are requiſite to maintain them in a thriving condition; and when corn is ſtunted, weeds will certainly grow predominant; and then follows the great charge of employing a number of weeders a long time, to the damage of the corn, and the impoveriſhing of the farmer. But the ſucceeding farmer, with his foot-pecked ſhare-plough, ploughed the ground (as I ſaid before) into very narrow thoroughs, by which he laid the land evener and better, than the firſt tenant did with his wheel fallow-plough; ſo that he ſowed

his

his grain in a fine, loose earth, that caused it to grow apace, and out-run the weeds, to his great advantage.

CHAP. II.
OF ROLLING.

THIS is an ancient *Virgilian* piece of husbandry, most necessary to be performed in *April*, both on *Vale* and *Chiltern* corn-grounds, whether they be barley, peas, or beans, &c. for the several reasons following. First, to secure them from the destructive slug. Secondly, to prevent the damage of long and violent droughts. Thirdly, to nourish the corn-roots; and Fourthly, to make them stand fast. First, to secure a crop of corn from the slug. The slug, or naked snail, though a small insect, is the most mischievous of all others to corn-crops, especially to that of peas; for these are the most natural and most delicious food of all the field kind to them, and therefore they remain feeding on them longer than any other, for the slug attacks the peas from their infant growth till they are in pod; and it is on this account that a farmer, in the random way of sowing peas, cannot be sure of a full crop of them, till they are past the slug's power of hurting them; which

keeps

keeps many farmers under a panic apprehenſion of the loſs of their pea-crops, or ſome part of them, by this ravenous inſect; for the rapine of this ſmall creature is committed chiefly in wet, warm weather, that frequently happens the month of *April*, which aſſiſts the ſlug not only in its vigorous feeding, but likewiſe in its breeding, by laying its ſmall eggs in little cells, under clots of earth; and when the weather is thus favourable to the breed of this ſlimy inſect, their increaſe in a ſmall field is infinite, and endures years together, if the ſeaſons are mild: that is to ſay, if a wet, warm ſummer and a mild winter ſucceed each other, then the ſlug remains in ſafety, and keeps his poſſeſſion till the next ſpring time, when he is ready for renewing his wonted rapine. But in caſe a very hard winter happens, then the farmer ſtands a hopeful chance of being delivered from thoſe his arch enemies, who, although they lie too deep for being deſtroyed by the tines of common harrows, yet a ſevere and long froſt will ſometimes penetrate into their beds, or cells, and there deſtroy both them and their eggs, as it happened to them in the winter 1739, and in the ſpring 1740, which was ſuch a hard and long froſt, that it rived, or ſplit, many oaken trees, or rather burſted them, by ſwelling their ſap into an icy ſubſtance, and at the ſame time, entered the earth ſo deep as to kill moſt of the ſnail or ſlug

Chap. II. OF ROLLING.

slug tribe, common worms, darrs, and cankerworms, infomuch that our crops of corn have hardly suffered by them to this day. But when these insects live in great numbers, in weather suitable to their nature, they become the greatest field pest of all living creatures, by annoying, and sometimes ruining corn, turnips, artificial grasses, and other vegetables, both in gardens and fields; and what very much chagrines the farmer is, that he cannot destroy them with all the art he is master of. Lime, indeed, will do a great deal towards it, but it will not fully answer his purpose; for, if to-day he sows his lime over a young crop of peas, turnips, &c. perhaps the same night or next day, great part of its efficacy may be lost by the fall of great rains: or, if a course of dry weather should succeed, then very likely the lime will serve to burn up the infant sprouting grain, as young turnips, clover, flax, wold, peas, and other green tender vegetables: and if he sows soot instead of lime, or tobacco-dust over them, he may chance to share the same fate; and so of any other hot powdered manure. Wherefore, to prevent these destructive mischiefs, as well as the farmer can, he makes use of the common wooden roll, eight feet long, fixed in a frame, and drawn by one or more horses: the hindmost being fastened in a pair of wooden shafts, for the more steady drawing, and turning it at the land's end, and bringing it through narrow gate-ways, that lead from one field to another.

Here we roll the corn that lies in broad-lands, and in ridges, commonly twice in a place, to crush down and kill the flug; and, for doing this work the more effectually, a good husbandman will begin to roll his ground at peep of day, for drawing the roll along the lands, while these insects are in their high feeding; for, if he defers this work later in the morning, the flug (especially if the weather is dry) will quit the surface, and creep into the lower earth for sheltering themselves, and lying safe from accidents. This caution, therefore, ought to be well regarded by all concerned in rolling of corn-crops, for one hour's time in the fore-part of the morning, is better than three afterwards, for killing flugs, by the pressure of the heavy wooden roll. Secondly, the roll ought to be made use of to prevent the damage of long droughts, which may prove fatal to crops of corn, by drying and parching their roots, while they are in their infancy; because then the sun and air has free access to them, and, being in their thready condition of growth, the heats may very easily enter, dry the earth about them, and do that mischief, which, perhaps, can never be recovered. Now, to prevent this, a good farmer, in due time, will roll his horse-beans, his thetches, his peas, his barley, or other crops, that

Chap. II. OF ROLLING.

that generally want rolling in this month, by crushing down, and laying even, the clotty earth on the surface; which, though it did some good before, by breaking off the gusts of cold winds from the corn-roots, yet now becomes necessary to break them down, by the pressure of the roll, for laying the surface level, for the operation of the scithe hereafter, and for shading and sheltering them from the power of long and dry hot seasons. Thirdly, by crushing down the clotty part of the surface with this wooden roll, and laying it even, the corn-roots receive a sort of second dressing; for, when the earth is thus squeezed about them, the rains will wash out its saline quality on them, nourish them all the summer after, and by this means assist a scanty manure, that was laid over the crop at sowing-time, and further increase the fertile effects of a rich one: as is obvious to all farmers, who practise the good husbandry of using the roll in this manner; and more especially so, if a dripping summer succeeds this profitable work, for then we seldom fail, by this means, to see our crops grow and flourish in large stalks, pods, and ears, that give us the greatest hopes of enjoying a plentiful harvest. The contrary effects of this management are easily perceived, when it so happens, that a farmer is put by his intention of rolling his corn in due time, as it now and then

then happens to be his case, by the fall of heavy and long rains, that hinder his doing this work; for if he was to roll his corn in such weather, the horses feet would be apt to stolch, crush down, and bruise, or bury many of the cornstalks, or blades, to the utter spoiling of some of them; therefore it must be dry weather, whenever this work is performed. But, if this happiness is not to be enjoyed before the stalk and blade are got too old for the purpose, then the rolling-part had far better be let alone, and the farmer remain content to see the want of that he cannot have; which, very probably, will cause him the sight of a languishing crop of corn. Fourthly and lastly, by rolling corn, the clotty surface of the earth is so fastened on, and about its roots, as to enable the stalks to stand erect, for receiving the benign benefits of the sun, air, and rain, in perfection; for, when they grow in this posture, the sun and air have a free passage between their many upright stalks, that thus serve to ripen them with the greater expedition, because the aerial, warm influences have room to harden them, and cause the stalks and ears to perform their regular and safe blossoming, and blooming, without which there can no right, full crops of corn be reasonably expected; for, when the stalks of corn are, by heavy rains, strong winds, and for want of rolling the ground, forced

Chap. II. OF ROLLING.

forced to fall down, fuch crops cannot enjoy a right bloffoming, or blooming; and then, I am fure, they cannot be well corned.

Again, there is this caution alfo to be obferved in the rolling of corn; where barley, or other grain, has not been rolled before fpindling time, in fhould, by no means, be rolled then; for, if this was to be done, it may, perhaps, bring on the lofs of almoft a whole crop of corn, becaufe, if its ftalks are got fo ftrong as to grow upright in a tolerable height, and the roller be then drawn over them, it will confequently fqueeze them to that degree as to make them bleed, as we call it; that is, it will bruife them fo as to force out part of their fap; and then, as the green ftalks lofe that vital part which fhould nourifh and carry on their growth, they muft be ftunted, if not killed. And although one would think this cafe needs no advice to caution againft it, yet I have known fome topping farmers commit this egregious miftake, to their great damage, without perceiving their folly, till it was too late, and fome not at all; for, being ignorant of the damage that fuch late rolling may caufe, when they fee a barley-crop ftunted, they are apt to impute it to a wrong caufe, and not to the right one, of the late rolling of their corn in its fpindling condition. I knew a yeoman, who was owner of a hundred a year, a fine farmer, that was brought up to husbandry business all his life, be guilty of this very mistake; who went on committing it year after year, till at last he had so much damage done, by rolling his barley at an unseasonable growth, as opened his eyes, caused him to perceive his error, and become his own monitor, as well as to other farmers; for he frankly owned he had acted wrong in rolling his barley (in particular) too late, even when it had arrived to its spindling condition.

On chalky, sandy and loamy grounds, we *Chiltern* farmers do not always employ the roll, when they are under a wheat-crop; it is seldom done, but when we apprehend such grounds may suffer by frosts, wets, winds and heats: in this case, many do roll their wheat-crops, whether it is sown in ridges, or in broad-lands, and this either in *October*, or *November*, or in *January*, *February*, or *March*; for the roll must never be employed, where the wheat is on a spindling stalk; if it is, it is six to one odds, but it spoils the crop, by bruising the stalks: no, if it is done at all, it must be done before the wheat is on the spindle; and when it is done, it may be drawn, the length way, over the ridges, or cross them, as the wetness or dryness of the ground indicates. So barley ground must be rolled with discretion; for, if a farmer should roll his barley, when it just appears above the earth, he may ruin his crop by it, as many have done,

Chap. II. OF ROLLING.

done, and do to this day, through their ignorance: for it is plain, that if a barley-field was to be rolled, when the barley is but juſt above ground, the clots of earth, that the roll would break, ſpread, and cruſh down on the infant barley-ſtalks, or ſpires, would ſmother and kill many of them, ſo that they never could ſhoot more; and then the ignorant farmer, that overlooks the cauſe, complains of the loſs of great part of his barley-crop, falſly imputing it to the deſtruction of froſts, chills of wets, or worms, &c. when the true original cauſe is his rolling the barley crop too ſoon, and ſmothering the young ſprouting ſpires or blades; wherefore, a prudential farmer will not roll his barley-crop, till it is two or three inches high, at leaſt: for then, if a clot break on ſuch a long blade, the root, by this time, has got ſuch hold of the earth, as to be able to repuſh out ſuch a blade again, and carry on its growth, with the greater vigour, by means of ſuch a timely rolling. But, even here, it requires a particular caution; a particular one I ſay indeed, for if a medium is not obſerved between the two extremes of rolling barley too ſoon or too late, the farmer may equally ſuffer on either account.

In *Hertfordſhire*, and in moſt *Chiltern* countries, the rolling of corn lands is ſo neceſſary, that though a farmer may plough his ground, as it ought to be ploughed, dress it, and sow it in a workman-like manner; yet, if after it is sown with corn, this operation of rolling the ground is not performed, he may lose great part of his crop for want of it: to prove which, I have first to say, that, where a chalky soil is sown with wheat, barley, oats, or pease, either in the two-bout-ridge form, or in broad-lands, this earth will be in such a loose texture of parts, as to let either the frost, the sun's heat, or the cold air, or the washing rains, too freely to the roots of the corn, and very likely do the corn crop a great damage. Now rolling such ground is the only remedy, to prevent these mischiefs; because the weight of the roll closes the chalky, porous surface, and so binds it, as to let these extremities of weather to the roots of the corn in a gradual degree; and yet, for all the farmer's precautions, and diligent endeavours on these accounts, the power of frosts, winds, and heats sometimes overcomes all his art, and hurts his crops: for example, in the great frost of 1739-40, all those chalky lands, that lay most exposed to the north and east winds, were so shattered by the frosts and winds, that the farmer had hardly his seed again, in return for what he had sown for a wheat crop; because the frosts, in the first place, loosened the top earth so much, as to give the winds an opportunity to blow it

from

Chap. II. OF ROLLING.

from many of the corn roots, and then the frosts got such an easy access to them, as to kill most of them; and this, notwithstanding the benefit that the rolling part communicated to the corn crop. If then rolling the corn ground will not secure the roots of the grain, from the power of extreme frosts, winds, heats, and rains; how much less will it be able to withstand these violences of weather, if such ground was not rolled and closed at all. Therefore, the farmer that does not roll his corn ground after sowing, will, in course, suffer a great loss; and the more, if these extremities of weather happen to be very violent. Thus, in sandy lands, wheat is the most exposed of all other grain, to the fury of frosts, winds, rains, and heats; because this, as well as chalky soils, is the loosest of earths. Indeed, where a sand is of such very loose parts, that it will not admit of the growth of wheat, the danger is not so much in *Lent* grain, by reason it is sowed in *March*, or *April*, and generally free of the damage of frosts: but, as wheat lies all the winter abroad, it often suffers, in these two sorts of soils, beyond all other grain. And the most of all, where a farmer sows these soils of chalky, or sandy loams, with wheat, without any dressing or rolling; for then the extremities of weather seldom fail of destroying most part of such a crop, either in

winter,

winter, by frosts, cold winds, and rains, or in summer, by the violent heats of a long dry time. In this case, dressing and rolling the land stand the farmer's great friend, because the dressing enables the wheat roots to take strong hold of the ground, and the rolling helps further to fasten the earth about them, and keep out these aerial violences. Therefore, whether it be a wheat, or *Lent*-crop, that is set on these soils, rolling is one main preservative of such a crop. And where the wheat is sown in two or threebout ridges, in these earths, the farmer, that he may the more effectually close the top and side surfaces of such ridges, draws his nine feet long wooden roll athwart or cross these ridges, whereby near three parts of the land in four are rolled. Some therefore roll their chalky, sandy, and loamy corn grounds, quickly after their being sown; others, not till *January*, *February*, *March*, or *April*, fearing the fatal effects of *March* winds, more than frosts; for, if these blow the earth from the roots, they are exposed not only to the frosts, that may happen in *March*, *April*, and *May*, but also, if they miss being hurt by them, the heats may still dry them, so as to make them produce a bad crop. For the prevention of which, in our *Chiltern* country of *Hertfordshire*, we seldom fail to roll our horsebean and peas crops, as well as barley and oat crops,

crops, after the corn-heads have shewn themselves above ground. When beans are about an inch or two high, we draw the roll over them to close the surface, and new mould up their stalks; which adds a fertility to their growth, as well as helps to keep out frosts, wets, and heats, from doing them much harm.

BOOK IV.

Of the CULTURE of WHEAT.

CHAPTER I.

OF PLOUGHING FOR WHEAT.

A Farmer, famed for his skill, lost many acres of wheat in 1740, by sowing it in stitches, or two bout lands, on a chalky soil, that lay to a northern aspect, which gave this wind, with the help of long frosts, an opportunity to crumble and blow away this loose earth from the wheat roots, and ruined the crop. However, as experience is the best schoolmaster, it taught him to sow his wheat, for 1741, in broad lands in such light land. When he was asked, Why he sowed his wheat in stitches in a chalk? he answered, he used to have good crops by it, but he found an alteration to his cost this year.

About the Hyde, there is a clay bottom, and a stiff loam at top. They fallow
with

Chap. I. OF PLOUGHING, &c. 157

with the fwing-plough, by laying two of their fize-lands into one, which then contain eight bouts. At the firft ftirree-time, they plough in the fame manner. At the third ploughing, they fill in thoroughs, and plough a-crofs into broad-lands; and if the ground is very four, they plough it a fecond time, then fill in the thoroughs, and harrow all plain; and, the fifth time, they plough and fow wheat in four-bout fize-lands; here they feldom fow any other grain than wheat and horfe-beans.

A ten-acre inclofed field afforded a difmal fight, being thought to contain no more wheat than would grow on one rood, or half an acre, of land well planted. The reafon was allowed to be, that this farmer gave the ground, which was a clay mixed with gravel, only one ploughing in *April*, and another in *September*; in which fpace of time, the knot, or couch-grafs, and other weeds had fo taken it, that the foil got very foul and four. Notwithftanding this, our bad hufbandman ploughed in his wheat-feed in four-thorough ftitches, or two-bout lands, that had been well dreffed all the fummer before with the fold, and in the fpring following with foot, but all was near loft for want of ploughing the land oftener, as was apparent by the feed's being bound in, and not able to put its blade through fuch tenacious glebe; which fhews the excellency

of

of the plough's use, whose business it is to reduce such surly soils into a powdered state, and make them give room to their fixed salts to act and nourish the vegetables: and so great is the power of this instrument, in this respect, that a piece of ground, well ploughed at several times, and got thereby into a fine tilth, shall, without any other help, bring forward a much better crop of wheat, than a half-ploughed four tilth double dressed, whereof this very example is a sufficient proof.

In order to get the ground in a readiness for sowing it with wheat, about the beginning of *October*, they fallow it up in *April*, and commonly by ploughing two of their four bout lands into one; at the first stirree-time, they plough this one bout land, and another such, into one broad-land; at the second stirree-time they hack it athwart, and next harrow all plain: then, at the fourth or last ploughing, they sprain and plough in their wheat-seed in three or four bout lands, the way the drain lies, leaving two water thoroughs by moulding up the earth out of them to the two bout lands, and little or no seed in them; and thus it is that they lay up their wet land high and dry, and leave, at the same time, a large passage for floods of water to run quickly off. But to be more particular in my account of improving these flat, four, heavy, and wet
earths,

earths, by a new method of ploughing, I have this notion to advance, that, in a dry feafon in fummer-time, two horfes may be put a-breaft and one or two length-ways before, to draw the fwing-plough; or, if they have an iron cock with notches at the end of the beam of their plough, or what fome call notch-geers, they may plough with horfes all in length a-crofs their broad-lands; that is, after they have ploughed three or four of their fmall lands together into one broad-land, and fo throughout the field, they may, after harrowing all plain, plough their ground a-crofs into fingle bouts, which is what may be faid to be a full half ploughing of the earth. The next time, it is to be bouted off the laft bouts; then run the plough through the middle of each bout, which will almoft level the whole land, and then it is all ploughed and broke; which, when harrowed, is fit to be ploughed a-crofs into three or four bout lands, and to have wheat feed fown in the fame. This is a much better way to get their ground into a fine fweet tilth than is commonly practifed at prefent, and exceeds all flat and other of their ridge-ploughings; for fo it is, that thefe farmers ftand in the greateft need of any item, that may be ferviceable to inftruct them, how they may get their ground fine and fweet, becaufe, without thefe qualifications, they have not much rea-

fon

son to hope for a full crop of wheat. But, instead of single bouting up their ground, they plough it, by way of hacking, once in a summer; which, indeed, is a sort of twice stirring or ploughing the earth, but it does not break and sweeten it, nor kill the weeds, like single bouting it. If these *Middlesex* farmers can enjoy a dry ploughing season, and plough their ground, after the manner I have directed, they need not be at the trouble of burning the couch-grass that their sour and clung fallow land produces, every year, in abundance in many places; nor likewise at the great charges of weeding their grounds, in the spring time, when they are forced to employ many hands to get the weeds out of their wheat, pease, barley and beans; not only to the great damage of the wheat by the tread of the weeders, but otherways, to their great loss, for I have known it cost one of their middling farmers ten pounds in one year to clear the fields of weeds. But this is not all that I have to observe on the negligence of many of these *Middlesex* farmers; there is another main fault that more or less of them, every year, justly deserve to be upbraided with; and that is, their fallowing their stiff clayey lands so late as in *May*, which is as great a mismanagement in these farmers, as it would be for the other sort of farmers, whose land is a rashy gravel, or a light sand, to plough it too soon; for one ploughing,

in

in such light earth, does generally as much service as two or three in a much stiffer soil; so great a difference is there in the knowledge of farming. For example, if I was to plough up my small pebbly, rashy, gravelly field the beginning of *April*, and plough it four or five times, the same summer, for sowing wheat; and also, if frequent rains should happen in that time, great part of that little mould that lies between the small stones, might, perhaps, be washed a considerable way down, to the impoverishment of the next succeeding crop of this valuable grain, because the bottom of this soil is of so loose a texture, as to give a very easy passage to rains, for carrying down much of its best earth. But it is not so with those stiffer lands, which lie so flat and wet, that to my certain knowledge, some of them could not be got into a fine and sweet tilth in a whole summer, though they were ploughed, early and late, five several times in that space. Nay, I may affirm with assurance, that many fields, even in our *Chiltern* country, whose soils were of the stiff sorts, but lay much more commodious for a quick drying than those in *Middlesex*, could not be got into a sweet and fine tilth in a whole summer, notwithstanding all the art and labour that were used to obtain the same; and then follow the losses which commonly attend the sowing of wheat-seed in rough twitchy, and sour tilth-

tilth-earths; as, the burying of a great deal of the feed, so as never to come up, the excessive thinness of a crop, a poor, small, dwindling ear and kernel, and the great charge of cutting out weeds from the corn, that such ground seldom, if ever, fails of being the occasion of. But the damage does not end here; I must also observe, that, when such a sour tilth produces a thin poor crop of wheat, the same ground is more than ordinarily exhausted by the growth of weeds, that infallibly grow up where corn does not. On the contrary, where a full crop of wheat is, there the weed is overcome, and, being prevented from growing, the bulk and cover of the crop keep in the spirit of the earth; because the sun, here, cannot exhale it, nor force the breed and growth of weeds, but will so hollow the ground, that one ploughing, next time, will prove as good, and go as far, as two or three would, if a thin, poor crop of wheat had grown on the same.

The person, I here write of, lives about a mile distant from my house, has the reputation of an excellent husbandman, and justly deserves this character in many respects; but, in this I am about mentioning, he does not, because he generally lays too much dung on his ground, for a wheat crop especially, which causes the land to be so 'ow, as not to be stiff enough to

keep

keep up the wheat, when it is in its green ear; and when it thus falls down, the fap is checked in its free afcent; and the kernels thereby prevented from arriving at a full maturity and bigness: which is a plain proof that land may be made too rich as well as too poor, for getting a full crop of wheat or other grain on it; for if a hungry, binding, gravelly, loamy foil (as this man's is) may be thus damaged by an indifcreet dreffing it with dung; intire rich loams may be fooner brought into fuch a prejudicial condition, by the fame ill management.

These lines, then, may juftly upbraid all thofe farmers, who feldom, or never, employ any other dreffing than dung, and are fo indifcreet, as to ufe it directly out of the farm yard, for nourifhing a wheat crop; as, to my knowledge, many do, both in *Vale* and *Chiltern* countries, and thereby incur the damage of fmutty or pepper-wheat, the growth of weeds, and the fournefs of their ground. But they are certainly the beft hufbandmen who firft clamp and rot their dung, and lay it on for improving a bean-crop, and not directly for a wheat-crop: where wheat is fown in *December*, and the weather is not too wet or fnowy, and the ground will admit of it, a flock of fheep, with a rack of wheat, or bean, pea, or thetch-ftraw, or hay, will do great fervice, by feeding them in the fold for their more plentiful dunging, and keeping them in health: but barley, or oat-straw, in particular, is to be avoided for this purpose; because these are apt to shed their kernels, and stain the wheat, by their taking root among it; and then it would be the farmer's great loss, who must sell a sack of such stained wheat, for a shilling or two less, than if it was clear from such soil. In short, by thus folding on new-sown wheat, it will cause it, in a great degree, to resist winds and wets, and very much preserve it against that loss, which generally happens to the farmer, when his wheat falls down, and is laid before harvest; force the worms to quit their living near the surface, and gnawing the roots of wheat; and, at the same time, by the tread of the sheep, it will be so fastened and closed, that the field fowls cannot come at it to devour the seed, nor the frosts to damage it. The third way is done by rotten, short, stable-dung, ploughed into the ground at the very last ploughing of all, when the wheat seed is sown; this is a good way in one respect, especially where the wheat is sown in two, three, or four bout-lands; because, by this way of ploughing and sowing, the plough mixes the dung and lays it so mixed with the earth, in a very regular manner, to the great advantage of the seed's growth, which here is kept warm by the dung, and very much defended

against

against the damage of great winds, hard frosts, and violent rains: but then there is this inconvenience attending it, the dung which naturally is of a hot and mouldy nature, is apt to heat, canker, and spoil or damage the wheat seed, to the breeding of smutty or pepper-wheat: yet, as prejudicial as this is, it is more practised by the *Hertfordshire* farmers, than any other I ever met with: accordingly, I believe, I may venture to affirm it for a truth, that we, in the western parts of *Hertfordshire*, suffer as much, or more, than farmers do in other countries, by this old pernicious custom of laying on our rotten dung, just before we plough and sow our ground with wheat. And this, in a great measure, I prove, by what I have been an eye-witness of, in other parts of *England*. In *Kent*, and *Essex*, to avoid this ill piece of husbandry, they mix their dung with mould, and small chalk, or lime, early in the summer, against wheat season; which gives them an opportunity of turning their one great heap, and mixing its fertile ingredients together more than once; so that, in time, they become incorporated into fine minute parts, free from that mouldy heat, which, as I said, dung alone is subject to. And, thus, they not only prevent these disasters, but make such manure go a great way further in the fertilizing their wheat-crop, because it mixes with almost all parts of the earth. Others think even this too great a hazard, and fear, if they lay their dung on their wheat-land the same year they sow it, that it will heat and canker their wheat seed, to the breeding of smutty and pepper wheat. And therefore they make a compost or a great heap with several materials, which they lay on their land, when they sow it with oats, barley, or beans; so that in the following year, when they sow the same ground with wheat, they are pretty sure of being delivered from these two misfortunes. This they do in some parts of *Essex*, near the town of *Barly*, in that part of *Hertfordshire* next *Essex*, and in several other places. In *Norfolk*, *Surry*, &c. they lime their wheat-land; in *Hampshire*, *Cheshire*, &c. they marle it; in *Wiltshire*, &c. they dress their wheat-ground with a white manure; in *Cambridgeshire*, &c. they lay on oil-cake powder, and in some parts of *Hertfordshire*, they make use of soot, horn-shavings, rags, sheeps-trotters, cows and hogs hair and hooves, &c. striving as much as they can, to dress their wheat-ground with any thing but dung, and laying that on their meadow, or barley, oat, or bean-land; by which they avoid the great damage that naked dung often brings their wheat-crop under. The fourth way is done, by laying long dung over wheat. This practice, as I said before, is chiefly done through

through neceffity, when the farmer has not had dung enough by him in the former part of the year to drefs his wheat-land with, and therefore is glad to do it now, with that he makes by his ftable, his hog-ftye, or his cow-houfe. If by the firft, it highly concerns his intereft to make his ftable-dung, as I have obferved, from the feed of wheat, bean, or pea-ftraw, or hay, and lay it under cover, that the rains may not wafh out its virtue; and then, when frofty weather happens that he can draw it on and fpread it over his wheat, it will do great fervice. My way, in this refpect of management, is thus; I lay fuch ftable-dung, both long and fhort together, under cover, as it is carried out of my ftable; and, as I increafe the heap, my boy throws the chamber-pot over every part of it, which very much helps to rot and enrich it, and caufe it to do great fervice to the wheat-crop at this time of the year. But to this great conveniency, I muft tack the account of an inconveniency attending the laying on of long dung on wheat; and that is, that fuch long dung fhould never be laid on wheat after *December*, becaufe, by the length of the ftraw, it will be apt to fmother the young blades of the wheat, and thus do more harm than good. I fhould have faid that the blades of wheat in *January*, or afterwards, commonly get fuch a length and bignefs, as to cover moft, or all the ground; and then, if such long dung is laid over them, it will be apt to make them look yellow, and perhaps kill many of them; but by laying it on before, in the infant growth of the wheat, it will not only greatly nourish it by the help of rains, which will wash out the fertile quality of the dung and stale, but afterwards the straw part will keep the wheat warm, and shelter it against the north and east winds, which sometimes kill or cripple thousands of acres of this golden vegetable, and also enable it to resist the chill of waters, that in many flat stiff soils, are very apt to damage the wheat crop. Lastly, there is an opportunity to dress wheat crops by the hand, in sowing over them soot, peat-ashes, oil-cake powder, or cows or hogs-hair, &c.

CHAP. II.

OF WHEAT ON CLOVER LAYS.

I Had a field that returned me two good crops of clover in two years; and in this month, I gave it only one ploughing with our wheel fallow plough into broad-lands, and after harrowing once in a place, longways, I sowed my wheat seed, and harrowed three bushels of it on each acre; which, as soon as done, I laid on

fifteen

Chap. II. ON CLOVER LAYS.

fifteen loads of dung on every acre, that made it grow and flourish well till *January*, when I fed it down with my sheep almost bare, but it quickly recovered, and was an admirable good crop; for the winter was so mild, and the spring so warm, that if I had not thus eaten it down, I suppose it might have been too rank, laid and spoiled that way; however, as the soil was a loam and a red clay, I ran a risk, for if some weather had happened, it might have kept down the wheat and given the weed room to have been master.

Another farmer, near *Dagnal* in *Bucks*, took this method. After the field had laid down two summers under cover, he gave it a ploughing, and harrowed in wheat, which lay till the twenty-first of *November* following, and then he sowed over every acre twenty bushels of soot, in order to force on a forward, large, timely head, for a winter subsistence for his sheep, and it answered accordingly; for he not only enjoyed a feeding crop, but at the same time gave the ground another dressing, which, in all, were three real ways of dressing the land; first with clover, next with soot, and last with the sheeps dung and stale. But this, as I said, is not to be ventured on in all soils. This was a loamy, chalky earth, that lay low and warm, and not apt to run into weeds, which brought on an exceeding good crop of wheat at harvest.

Another

Another farmer, near me, fed his wheat down with his sheep in *January* and *February*, but lost most of his crop by it, notwithstanding his ground was in good heart, and dressed well. The reason was, that this man's soil was a wettish loam, that lay high, and was exposed much to the cold winds, that so crippled the short bitten wheat, as to give room for the black bennet and other weeds to get the dominion of the corn. The former dressed his clover lay on the top, this pressed his dung in, and indeed, it is thought by many, that top dressing is best in these cases, provided it be laid on immediately after the corn is sown, because it keeps the roots warm, and secures them the better from the frosts, than if ploughed in.

CHAP. III.
OF WHEAT ON SAINFOINE.

First Case. A Considerable *Chiltern* farmer having enjoyed a sainfoine crop many years, after it was worn out, he ploughed it up in winter by the help of gift-ploughs, and afterwards several times himself, till he had killed all grass weeds and roots, and got the ground into a fine tilth; then, about *Michaelmas*, he ploughed and sowed the same with wheat-seed, but

Chap. III. ON SAINFOINE.

but had not a quarter of a crop in return; however, to recover this his miftake, the next fpring-time he ploughed up the fame land, and fowed it with peafe. After the peafe were off, he ploughed and fowed the field with wheat again, and then had a good crop.

Second Cafe. Two farmers, living at *Great Gaddefden,* about two miles diftant from the former, hearing of the firft man's lofs, forbore to fow fuch ground with wheat the firft time, but inftead thereof, after having ploughed their fainfoine land feveral times, to get it into a fine tilth, fowed it the firft time with turnips, and after them barley, and then wheat, to great profit.

Third Cafe. Alfo farmer *Butler,* of *Wards,* near *Ivingboe,* fell into the like miftake, by fowing wheat for the firft grain on new broken up chalky fainfoine ground, and had hardly any in return. After that, on one ploughing, he fowed oats, and next time wheat, which then anfwered to his fatisfaction.

Cafe the Fourth. Nowithftanding the firft three cafes happened within two or three miles of another great farmer near me, yet he fowed his fainfoine new broken up ground the firft time with wheat in 1741, but had a poor crop for his pains.

Cafe the Fifth. However, a farmer near this laft would not go on fo, for he took a better method,

method, thus: about *Allhollantide* he began to fallow his worn out sainfoine ground, and, after he had by two ploughings more got it into a tilth, he sowed it with turnips, in *July*, 1742; for, he said, he had tried to get a crop of oats, after he had ploughed his sainfoine ground more than once, but could not come by a tolerable one; therefore he sows such land the first time with turnips, then barley, and then directly with wheat: for that his chalky earth raised so loose at first, that the corn-roots could not get a close lodging, which occasioned the misfortune. But, by sowing turnips the first, the sheep trod and closed the ground so well, as to produce corn in abundance: for such long rested earth is not only an enemy to a corn crop, the first time of sowing, by being in a very loose condition; but the worms also, that bred in its undisturbed pores for some years together, often do great damage, unless a turnip crop precedes the corn crop.

CHAP. IV.

OF WHEAT ON NATURAL GRASS.

THERE is another way of managing a lay of natural grass, that is to be done by drawing a foot-plough with a broad share on it, which will pare and turn up the turf very thin,

Chap. IV. NATURAL GRASS. 173

thin, much better than a pecked share wheel-plough: then immediately after this another foot plough may follow, with either a pecked or chizzel share, to turn up and throw fresh virgin mould over the turf, and so on till the whole field is finished, and then all will appear like one intire surface of mould; when this is done harrow it plain, and sow three bushels of wheat over one that is to be harrowed in.—Now, it may be wondered at, why I direct the sowing so much on one acre of ground as three bushels of seed, but this is easily answered; for, unless this quantity is sown, the crop may suffer, because in such ground there is generally not only a stock of worms and grubs bred by the turfs lying many years undisturbed, but, by the shallowness of the staple, the seed is more than ordinarily exposed to the beaks of field-fowl, and other devouring vermin, which may eat up so much as to leave a very few of the kernels to grow into a crop, unless a quantity of seed be sown accordingly.

CHAP. V.
OF THE SORT OF WHEAT.
SECT. I.
Red Lammas.

AS wheat is the king of grain, so this sort has been deemed, hitherto, the king of wheats, for having deservedly been under the reputation of producing the whitest and finest of flour, as well is known in particular to the *Londoners*, who call it, the *Hertfordshire White*; not that the best *Lammas* wheat grows in this county, but because of the vast quantities of this largest bodied sort, that grow in *Northamptonshire*, *Bedfordshire*, and *Buckinghamshire*, and some other northern counties, which are weekly brought up and sold at St. *Alban*'s, *Hempstead*, and some few other markets in these parts, where the greatest number of water mills are situated, for grinding corn in a lesser compass of ground than any where else in *England*. It has a red straw, a red ear, and a red kernel; and it makes the whitest of flour. However, this is certain, that red *Lammas* has not only a whiter flour, but a much softer, and finer, and often a bigger body than others; which makes it the most agreeable sort for the greatest quality, and therefore is the fittest wheat to sow on the best land, on account

of

Chap. V. OF WHEAT.

of its fetching the best price at market. This sort of wheat is likewise in great esteem with the *Middlesex* farmer, for its stiff jointed, high coloured, long straw, that they sell in great quantities to the *Londoners*, for littering their horses with it, sometimes at eight pence a truss in scarce times, that weigh about thirty-five pounds, which enables them to pay the greatest part of their rent with this one commodity. It affects to grow in the richest vale lands, consisting of loams, or black, or blue clays, where I have seen it grow to near five feet in height; yet it is by many sown in *Chiltern* dry loams, and even in some gravels, that have been before dressed or manured extraordinary well. In short, this wheat was, till very lately, the only sort sown in vales, and is that noble sort, whose kernels are somewhat longer than pirky wheat, and near as big as cherry-stones, when it is sown as true seed in a right soil; then it will outweigh the pirks or white wheat; and for these benefits, great quantities of its seed are bought and carried in waggons into the northern countries, because, about me and *Dunstable*, it is sowed in chalky ground, partly for the purpose of selling the seed to be sown in heavy loams, or stiff clay lands in those counties; chiefly for the sake of its change of soil, and because we get the same first out of the hands of those farmers who make it their

business,

business, in several parts of *Buckinghamshire*, to lay down ground every year, purely to improve this grain for its seed; for here it will come off in a large body thoroughly clean, and free from the seeds of all weeds. But this wheat has a long lank ear, that exposes it more than all others to the spoil of blights and strokes; for which reason a right manager will sow this wheat the first of all others, that it may get into an early hard ear in the blightening honey-dew seasons of *June* and *July*. Therefore sow this wheat in *September*, or the beginning of *October* at farthest. And it is on this very account, that many farmers of late have acted more judiciously than formerly; for, in order to avoid this fatal misfortune, some have sown it early with a mixture of pirks or white wheat, whose ears, growing in a thick bunchy shape, prove a sort of skreen, or shelter, to keep off and break the honey-dew, and other blightening causes, from the red *Lammas*, in a great degree. Red *Lammas* wheat, when thorough dry, will weigh six bushels weight; I mean, five bushels in measure will weigh six bushels in weight, yet will not yield quite so much flour as the same quantity of pirky wheat will. Red *Lammas* wheat grows about six or twelve inches higher than pirky wheat; and as it shades too much, it is not so proper to sow with pirks, as some think, though others think otherwise.

SECT. II.

Pirky Wheat.

THIS wheat has a white straw, a white ear, and a red or yellowish kernel, more round than *Lammas*, and has more of late, than ever, got into reputation for its profitable qualities, particularly for that one of growing in *Chiltern*, gravelly, and chalky soils, where it will flourish and yield excellent crops, when the red *Lammas* will in some measure fail, because this sort of wheat will grow in a poorer soil than that, and yet return as many bushels as *Lammas* will in rich loams or clay, provided such gravels, chalks, and light loams, are well sowed and dressed with a sufficient quantity of manure. Its bunchy ear is not so subject to be blighted as the red *Lammas*, because its corns grow close in a shorter ear. The kernels of pirky wheat have rather a thinner skin than that, and, if of the right sort, will yield abundance of fine flour, and the more when they are full dry, for then they will crumble under the stones into much meal and little bran; and as the very finest and shortest bran cannot be easily separated from its flour, it thus becomes coarser than that made from *Lammas*, but this is compensated by the quantity it makes. For this reason, when pirky wheat is got into the barn dry, and has a good body and colour, it will fetch at market near, if not altogether as much as *Lammas*.

This pirky wheat has but a single chaff, so that when it is ripe, the kernel may be almost seen through it, for then it is ready to start out of it. Five bushels of pirky wheat seldom weigh more than five bushels and a half in weight, or three pecks at most; yet if the same measure of the red *Lammas* is ground, the flour of the pirks will weigh more than the flour of the red *Lammas*, for the foregoing reasons.

Pirky wheat has four or five sets of kernels, when the *Lammas* has but two or three. It also has this good quality, not to draw the ground so much as the *Lammas* sorts, because the straw of the *Lammas* is taller and bigger than that of pirks. When pirky wheat is sown in gravelly ground, it generally acquires such a colour, that it often deceives the buyer for *Lammas*, and is accounted the thinnest skinned wheat of all others. The red pirks are best, the yellow are rather thicker skinned, and hardly to be known from yellow *Lammas*, as the red sort is from red *Lammas*, and the white pirk from the white *Lammas* wheat. In gravels and lean loams, the pirky wheat and *Lammas* grow much of an equal tallness when mixed.

SECT. III.

Yellow Lammas.

THIS wheat has a white straw and a red ear; its kernel is of a yellowish red colour, rounder, shorter, and not so guttery as the red *Lammas*; nor is the ear of the yellow so long by an inch as that of the red, is nearer of the pirky nature, has a flour near as white and as fine as that of the red *Lammas*, and whiter than that of pirks and white wheat; its skin is a little thicker than that of the red, and thus it runs something more into bran. The flour men do not care to buy this yellow wheat in summer, by reason its thick skin is then so dry, as makes it apt to grind into a fine bran, that mixes so much with the flour as to make it coarse. But in winter they readily purchase it, for then its skin quits its flour with ease. This wheat will grow on chalks, gravels, clays, and other poor land, much better than red *Lammas*, but degenerates sooner than most other wheats; for you may in a few years sow it too long in the same soil, till it dwindles to so small a kernel, as not to be worth sowing. Therefore, it is now very common to sow it in a mixture with pirk, because it is a much surer crop this way than when sown alone. Yellow *Lammas* yields a peck or two of flour less, in five bushels,

bushels, than red *Lammas*; yet I know more than one farmer that sows yellow *Lammas* in their stiff clayey loams in the *Chiltern*, and thinks it the very best sort, for its thin skin, close ear, and good yield; and best agrees to be sown with pirks, because it grows about four inches higher, and shelters that the better from blights.

As a *Chiltern* farmer cannot well pay his rent without crops of turnips or rapes, some, if not all his turnips or rapes, cannot be eaten off time enough for sowing the same land with a *Lammas* wheat, because this wheat growing on a long strong straw, and in a long ear, where its kernels stand the furthest apart of all other wheats, requires an early sowing to ripen it in due time; therefore, after *October* we account it rather too late to sow this sort, lest the farmer lose great part of the crop by it, because, by such late sowing of this *Lammas* wheat, it will very probably be too late in its green ear, and thereby be the more exposed to the damage of honey-dews and blights, for this wheat, more than any other, is least capable to resist these sort of prejudices. On this account we have recourse to the great conveniency of sowing pirky wheat, by reason this has a shorter straw and closer ear; is a hardier sort than the *Lammas*, for by its low slender straw, and bunchy ear, it is the better able to resist honey-dews and blights. Again, as turnips
are

are generally sown on a gravelly, a chalky, sandy, or dry loam, and eaten off by sheep or bullocks; the *Lammas* wheat will not agree with such dry lean earth, near so well as a pirky sort; for a loam best affects a rich stiff soil, to maintain its long thick straw and lank ear, when a pirky wheat will flourish on a leaner and dry one, which makes it of great value to follow a crop of turnips; and on this account it is, that all over *Hertfordshire* its farmers make use of hardly any other sort to sow after these roots, and this very late in the year, for I have sown it in the beginning of *March* upon necessity, and as it happened, I had a good crop late in harvest.

SECT. IV.

White Wheat.

THIS white wheat has a white straw, and a white bunchy ear; it kernels as big as pirks, grow closer together than the red or yellow *Lammas*, and thus is better secured from strokes and blights than they are. Now this corn is mostly defended against these incidents by means of its rough chaffs, for the right sort of white wheat has two or three that encompass each kernel, which also preserve it in a great measure from the damage of flies and other insects, which are apt to spoil wheat while it is growing in the ear.

When this excellent wheat is near ripe, its true sort may be known by its aspect, for then it will appear as if it were hoary all over its ear.

This wheat weighs rather lighter than either the red or yellow *Lammas* or pirks, yet makes more stuff or flour than they do, because its very small bran remaining among it is not easily perceived, for as its coat or skin is of a light colour, it gives the meal and bread a pleasing yellowish cast.

There is a smooth-eared sort of white wheat, that with us is not reckoned so valuable as this rough eared sort, and are both distinguished by the names of white pirks and white *Lammas*. But I must own, I never saw such large kernels of white wheat as I did in half tubs, as they stood in *Taunton* market in the year 1737, where the farmer sells this, and other sorts, by the peck and half-peck to the weavers and country poor people, who here call it *Holland* wheat; and in some markets it sells for more money than any other: but at *Hempstead* it sells for one or two shillings less in a sack than the red *Lammas*, or pirks, because here they chiefly endeavour after getting the finest flour for supplying the *London* baker and pastry-cook. The flour of white wheat is of so short a nature, that the workman can hardly make a loaf of it that will stand without cracking; and therefore it is generally mixed

Chap. V.   OF WHEAT.   183

mixed with the flour of the *Lammas* sort, which improves each other, so as to make excellent bread. White wheat will grow well both in vale, swampy, stiff soils, and in *Chiltern* dry grounds.

SECT. V.

Dugdale Wheat.

IN *Essex* they call this *Grey-poll Rivet*; in *Huntingtonshire, Dunover Wheat*; in the west-country, *Grey-poll* and *Blue-poll Wheat*; in *Hertfordshire*, Duck-bill, or Dugdale *Wheat*. In *Berkshire*, there is a white-cone wheat, whose straw is like a rush, full of pith, and not hollow, like the *Lammas* sort; however, these are all bearded wheats. The common duck-bill wheat has a darkish, brown, crooked, guttery kernel, rather bigger than any other wheat; but its chaff is better thrown away to make dung, than to give to cattle, because, by the sharpness of its sides, it pricks and cankers the horses mouths. It should be sown a fortnight before *Michaelmas*, at latest, because it is always late ripe, and then in a stiff soil; for this bulky wheat is very apt to fall by the wind and rain; therefore no light sand suits it.

This seed, when sown in a fine, well dressed tilth-ground, produces the most of all others.

One ear has been said to have had above an hundred grains in it, and on an acre above fifty bushels of wheat have grown; but, as its flour is of the coarsest and heaviest sort, it is sold for two or three shillings less in a sack than the *Lammas*, pirks, or white wheat. Its kernels are of a brittle nature, and therefore grind well with the *Lammas* sort. One bushel of *Dugdale* and three of *Lammas* make very good bread; but in many places in the west-country, the bakers in general allow more of this flour than the other, which causes their common loaves to be brown, close, and heavy, as I observed (and that most of all) in *Somersetshire*, where, I believe, they sow more of the grey and blue-poll wheat, than in any other part of *England*. In *Hertfordshire*, *Bedfordshire*, *Bucks*, and in most of the southern and eastern counties, little of this wheat is sown, and then it is chiefly for the farmers use, because a great deal of this will grow in a little rich ground, which enables them to make the most at market of their better sort. This wheat is fit to reap when its beards are ready to drop off; are hard to thrash out, and yield a great deal of chaff.

They never brine or lime the seed of bearded wheat, for it is never known to smut. Two bushels of seed is the common allowance for one acre of ground, as it is a great brancher. It

will

will grow in rough four tilths the best of any wheat, therefore some sow it on only one ploughing up of a lay, and harrow it in. One bushel of this mixed with flour of *Lammas*, the *London* baker approves of, because it keeps the bread a day or two longer moist than usual, and yet its natural coarseness is not perceived by the customer.

SECT. VI.

Egg-shell, or Mouse-dun Wheat.

I Do not know that any of these grow in *Hertfordshire*, or any where else in *England*, but in one particular county that I have travelled through. Here they told me that these are synonimous names for this wheat, which they affirm to be the best in *England*, because it makes the finest of flour; and, indeed, I was surprized to hear afterwards one of the greatest bakers in *London* confirm the same, who assured me it beat our red *Lammas* for making a white bread, that the flour of red *Lammas* could not come up to. I have handled some of this wheat, and must own, by its outward appearance, it did not promise all this; but its inside was, by all in the neighbourhood, allowed to answer the character I have given of it.

SECT. VII.

Change of Sorts.

IT is yearly experimentally proved by many curious farmers, that the change of feed gives a great improvement to a crop of grain, and especially to wheat; which being the chief corn that meets with the most certain sale of all others, and on which, as I said, two years rent depends, calls for the best management that can be given it. I knew a great farmer so careful in this matter, that where he sowed his red *Lammas* wheat one season, he sowed a yellow *Lammas* the next season; and so nice was this man, that he carried on the like piece of œconomy to the third season, or ninth year, by sowing white wheat on the same ground; for it was his notion, and what he has truly experienced, that where even pease were sown twice in the same field, though a fallow and a wheat crop came between, the latter would prove much the worse crop for it. Another of my acquaintance, who rents a large farm between *Hempstead* and *St. Albans*, sows five sorts of wheat in his inclosed fields; as the red and yellow *Lammas*, pirks, *Dugdale*, and white wheat, and assigns this reason for it; that if one sort missed he had the greater chance for another to hit, and thereby became the better enabled to change his seed, and adapt a proper

sort

sort to a proper soil. From whence I observe, that seeds of vegetables, like animals, affect to be nourished by variety of food.

This knowledge induced a gentleman, who lived near *Acton*, in *Middlesex*, to commission me to buy him forty bushels of the right sort of pirky wheat, that grew last on a chalky soil near me, to sow it in his low, wettish, stiff, loamy, clayey land; and he not only had a most plentiful return, but such a bodied wheat as beat all others in *Uxbridge* market, where he always sold his grain. Near *Amersham*, in *Buckinghamshire*, a farmer in the spring, 1741, having sowed some of his own bean-seed in one part of his common field, that he had before sowed with the same several years, at last was persuaded by a friend to buy some horse-beans at market for change-sake, and sow in the other part of the same field; and the last sown proved by far the superior crop. The rath-ripe barley, if sown more than two years, with us, will degenerate into our common barley, notwithstanding our sowing it in warm dry soils, as the gravels and chalks are; for this, at first, comes from off a sandy loam, about *Fulham*, *Hammersmith*, and *Chelsea*.

If this year we sow that barley-seed which came off a chalk, or gravel, on our red clays, or wet stiff loams, the kernels will be of a reddish colour and thick skinned. On the contrary, sow

the

the fame feed in their former foils, and they will become white bodied and thin skinned, which plainly shews, that it is not so much the change of climate that occasions such alteration, as the soil.

For a further proof of this, the *Aylesbury* vale farmer (whom I take to be one of the most obstinate bigotted sort) has at last been prevailed on about twelve years since, to sow barley now and then, instead of wheat, on his ridge lands by way of change, and, since, has found so much advantage by it, that the wheat, barley, and bean crops are all greatly improved by such a change.—The *Acton*, *Edgware*, and *Uxbridge* farmers are also, for the most part, under the custom of sowing only wheat and horse beans: but, if they would sow barley between, they would certainly find their account in it, for more reasons than one.—If white wheat is not changed from one soil to another, it will degenerate, and grow into a steely corn, of an ugly dark colour, little better than *Dugdale* wheat.

A great farmer, near *Beech-Wood Park*, by *Market Street*, used to sow a hilly field with wheat, as far as the hilly part went, because it was a stiff clayey soil; but, the bottom part being a gravelly stony piece of land, he always in the season sowed it with barley for many years,

till

till at laſt he was perſuaded to ſow all of it with wheat, and there grew a moſt admired, ſtrong, reedy ſtraw, and the beſt of wheat, which was imputed chiefly to the change of ſeed.—The yellow *Lammas* wheat, in particular in the whitiſh loams about *Ivinghoe Aſſon*, has been obſerved to decline, if you ſow the ſame ſeed in only two ſucceſſive ſeaſons, on the ground it was ſown on before: for it is their received opinion here, that they may ſow this ſort (as they term it) till they loſe it; that is, till it degenerates into a moſt poor thin kernel. Another great farmer that I was acquainted with, uſed to keep that wheat two years that he ſowed for ſeed; and, as it grew on hilly chalky loams, the vale waggoners would buy and carry it down to be ſown in their bluiſh clays, and black moiſt loams, for the *Lammas*, or pirks, coming off ſuch dry loams, proved an excellent change; and by means of its age, the ſmall underling kernels would be hindered growing, for in that time, they had loſt moſt of their radical moiſture and property, and accordingly they ſuppoſe the reſt of the larger kernels will never grow into ſmut. Another inſtance I bring, for proving the great benefit of changing ſeed every year, is this: I know a curious gardener, who ſows his broad beans in drills, made by the foot-plough, and covers them by hand houghs. His ſeed he carefully ſaves, and every year car-

ries

ries it twenty-two miles to *London*, where he sells it at the seed shop, and, as his beans come off a gravelly clayish loam, he gives money to boot (if his is not so good) for *Windsor* beans that come off a dryer soil. But Mr. *Tull* lays such stress on houghing of corn, that he writes there is no occasion for changing of seed if so managed.

A farmer at *Hudnall*, near me, who rents about a hundred pounds a year, is so happy as to enjoy diversity of soils in the same. One field has a clay bottom, another a loam, another a gravel, and another a chalk. These give him an opportunity every year to change his wheat seed. Now this man sows only two sorts, the red *Lammas*, and the pirks; if he sows the *Lammas* on his clay this year, next he sows the same seed on his gravel, or chalk, which though these last are not truly proper soils for this sort of wheat, yet this proves no impediment with him, for he dungs or manures the same ground so well, that he seldom fails of a good crop. In the same manner he uses his pirky wheat, which is a sort that grows well in any of the four soils. But this is a conveniency which few enjoy; therefore, the next way to supply it, is to change seed with neighbours who have different earths; but where such a correspondence does not reign, the market may supply it, that affords a variety at most times.

times. Yet this farmer did not trust altogether to the change of his soils for keeping his wheat-seed from degenerating; he had also a great dependance on the preparing his seed for sowing, after a different way from all others, that it might the better answer this purpose.

CHAP. VI.
OF THE TIME OF SOWING.

IN the *Vale* of *Aylesbury*, they seldom or never begin sowing wheat till *Michaelmas*, lest their fruitful soil cause it to grow rank and winter-proud, and to spend itself in growth, as to appear, in *April* and *May*, of a sickish, yellow, dying colour, when it then should shew itself in a most flourishing, verdant green state and lively condition. After a ridge half acre land has been well dunged or folded over, or otherways well dressed and ploughed into a fine tilth, they harrow all plain, and sow it broad-cast twice in a place, by crossing the throw. Thus the seedsman steps backward and forward, in all four times, till he has done sowing the half acre land, and then begins to plough all the seed in with the foot-plough by the earth.

It concerns every farmer to avoid sowing in *December*, unless there is such a prospect, or
other

other necessitous reasons; I say, a necessity that may thus happen. In case the turnips, or rapes, are eaten off, and the ground is ready to be ploughed and sown with wheat in dry and open weather in this month, then it may encourage the ploughing and sowing such a field with wheat; because, at this time of the year, this sort of weather seldom happens, else the sowing of wheat ought to be deferred till next month; for this is a standing rule with many of the experienced and knowing farmers; that wheat sown in a favourable season in *January*, will be as forward as that in *December* before *Christmas*.

An old sagacious *Chiltern* farmer of my acquaintance being asked, why he did not sow his ground with wheat in *December*, when it was prepared and made ready for it, answered, that a fortnight or three weeks hence would be much better, because the days would be longer, and the weather come on warmer; for, if wheat was sown in *December*, or even in the beginning of *January*, the frosts would be apt to overtake it on the sprout or chip, and very likely spoil it. Therefore it would be much safer to sow it towards the latter end of this month, and then it will be as forward as that sown even in the beginning of *December:* hence I would observe, that, in case wheat-seed is not sown in *September*, *October*, or *November*, it is better to stay till

January,

Chap. VI. OF SOWING.

January, for the foregoing reasons, as well as those that follow; for by this time such wheat makes its first sprout: in *February* the sun's heat increasing in a considerable degree, it will, unless the frosts are very severe indeed, come on apace. Yet none, I presume, will be so silly, as to refuse to sow wheat in *September*, *October*, and *November*, where they can conveniently do it, for the sake of sowing it in *December*, or *January*; for at best all those, who sow wheat so late, run a great risque of getting a plentiful crop; for the late sown wheat is liable to several disadvantages. One is, that the frost may keep it so long in the ground before it vegetates or sprouts, that it may rot.

Secondly, If it has a good sprouting or chipping-time, it may be yet killed by the frost. *Thirdly*, By late sowing, it will be late in bloom; and then the hot, dry season, may dry the green kernels so much, as to make them become no bigger than oat-meal grouts. *Fourthly*, The honey-dews, that fall most in *June* and *July*, will be apt to close and glew up the green ear so tight, that the kernels cannot enlarge themselves, or so spot and poison the straw as to hinder the free ascent of the sap, and then ensues a sort of blight. *Last of all*, Late sown wheat, by the shortness of the days, and the length of the nights, and the rains that sometimes fall in the latter end of *August*, and in *September*, often cause the farmers a difficulty to get the wheat home in thorough dry order.

Wheat is seldom sown so late as *November* in the *Middlesex* low wet grounds, nor in the vale of *Aylesbury*, and others, where they lie so wet and flat, that they are obliged to sow their wheat in ridge lands; I mean in rood, half-acre, and acre-lands, that are always ploughed one way; but where turnips are sown, it is the yearly practice of many farmers to sow the same ground with wheat in *November*, *December*, *January*, or *February*, and sometimes after cole or rape is eaten off, &c. But to be more particular, I shall mention several reasons for sowing wheat late: *First*, In case a farmer has a hot, rank, gravelly, or sandy soil, or a warm situated rich loam, he may be timorous of sowing such land with wheat in *September*, or *October*, lest it grow winter-proud, and so rank as to spend itself in too forward a growth before the next spring; for, if it should do so, there will then consequently ensue a poor crop. *Secondly*, He may be forced to sow his wheat late, if he be resolved to sow it on a clover, or sainfoine lay, that he could not conveniently get reduced into a fine tilth before *November*, *December*, *January*, or *February*. *Thirdly*, It may, and frequently is the case of many *Chiltern* farmers, to sow part of his wheat crop,

Chap. VI.   OF SOWING.   195

crop, after his turnips or cole are eaten off, which is seldom done till some of these months. And, *lastly*, Where a sward-ground is ploughed up to sow wheat, or where wood has been grubbed up for the same purpose, then, where a person is resolved to sow wheat as a first crop on the same, it should not be done till late, lest such a crop grow too rank too soon, be laid flat, and rot on the ground.

February is the latest that wheat may be sown in; but when it is to be done, a due regard should be had to the nature of the soil: If it is a stiff wettish earth, then two or three bout stitches or ridges, are the best form to sow the seed in, either by a single or double plough; and if a fold could be run over it after sowing, it might do great service to the wheat crop; because as turnip, rape, or rye-ground commonly turns up clotty, after these vegetables are eaten off, the sheep, by the hardness of their tread while penned, will very much break it, and make it somewhat finer than when the plough left it; will close the earth about the body of the seed, and prove a great security and fertility to it.

But where the land is of a dry short nature, wheat-seed is best sown in broad-lands in *January*, because, besides the cover of the harrows, oats, pease, beans and barley, being in many places sown in *February*, will employ the beaks of field fowls, and in a greater degree divert their search and feeding on the new sown wheat. But it would be a much greater security to all wheat-seed, sown in the broad-land mode, if a fold could be employed over it, and begun as soon as the seed is in the ground; however, as few can do this at this time of the year, if good rotten dung is immediately, in a thick quantity, spread over the wheat-seed, as soon as the harrows have done their work, it would not only be a security against frosts and chills of wet, and drought, but prodigiously bring on a forward and luxuriant growth, to the great advantage of the farmer; and the more, because such a fertile penning of sheep, or thick coat of rotten dung, will cause the wheat to get so quick a growth, as very likely to miss the strokes of mildews, and other blightening causes, which late sown wheat seed not thoroughly dressed, or manured, is most liable to suffer; for, when any wheat gets later than ordinary into its green ear, its stalks and ears are generally damaged by these incidents, oftentimes to the destruction of great part of the wheat-crop.

By sowing wheat-seed in *February*, as the days are got to a fine length, and the sun's heat increaseth on our island, the wheat-seed, sown in that month, will presently be forwarded in its sprouting; and, where it has a sufficient strength
of

of dressing or manure allowed it, it will make expeditious advances in its growth, and be thereby enabled to withstand those extremes of frosts, wets, and winds, which, in *March* especially, oftentimes kill late sown wheat in poor lands. But, to speak closer to the matter, I must say, the black frosts of *March*, and the severity of the northerly and easterly winds, which usually blow for a long time at this season of the year, have sometimes deprived farmers of near half the crop of wheat they would otherwise have, even of that which was sown early in well-dressed ground.

Much more, then, ought these fatal accidents to be guarded against in late sown wheat, whose infant blade and tender root are not so capable to withstand the violent chills of inundations of waters, long frosts, and cutting winds, as the more forward sown and stronger wheat-root and blade are. Hence it sometimes happens, that wheat-seed, sown in *December* or *January*, is frequently great part of it spoiled; because the seeds are so locked up in the earth by frosts, or chilled by wets for a long time, that it is often either damaged by stagnation, or killed by rotting in the ground.

It is therefore of great importance to all *Chiltern* farmers, who think it their greatest interest to sow wheat in *February* after turnips, rapes, or rye, to apply a strong and potent dressing, or manure, to their wheat-seed immediately after sowing; and the dressing, or manure, that soonest affects and reaches the seed, must assuredly be of the greatest security, and do the most service.

CHAP. VII.
OF THE QUANTITY OF SEED.

IN the blueish clays of *Aylesbury* vale, their rule is to begin sowing at *Michaelmas*, and then they sow something more than three pecks of wheat on a half acre of land, broad-cast, twice over the same, and plough it in with their foot plough. But if they sow a land late in *October*, they sow four pecks; and if they sow the half acre in *November*, any thing late, they sow five pecks on that quantity of ground. In our reddish loamy clays, about *Gaddesden*, we strive to sow our first wheat-seed; and if this happens to be in *September*, two bushels on an acre sometimes prove sufficient; if in *October*, two and a half; if latish in *November*, three are but enough, and then we sow it by spraining the seed out of a man's hand in two bout lands.

If the land is in a fine tilth, in good heart, and sowed about *Michaelmas*, the quantity of seed is three pecks; but if the half acre is sowed late,

Chap. VIII. OF SEED.

late, then more, but seldom exceeds a bushel. After sowing they never harrow here, but as they leave it they let the ground lie in the same posture all the remaining part of the year till harvest, for the top of this land will shoal and run into a fine hollowness even by very small frosts.

The wheat-seed that is sown in *January*, must be more than when it is sown in *September* or *October*, because, by being sown so late, it will not gather nor branch like that sown sooner, when a peck less upon each acre of ground was sufficient. Now, therefore, two bushels and a half, or something more, should be the quantity sown on an acre.

On poor ground more seed should be sown than on rich; because rich ground will cause most, or all the seed to grow, and very much branch, when poor ground will cause some of it to die, and little to gather and branch.

CHAP. VIII.

OF THE GROWTH OF WHEAT.

WHEN the wheat-stalk is well jointed, and is thick and strong, it is a sign of a good ear.—A small ear, and large grain; for the smaller the ear, the larger the wheat, rye, or barley.—A good cherry year, a good wheat year.—About two months after the shooting of the wheat-ear, harvest commonly begins.

<blockquote>A dry *March*, a wet *April*, a dry *May* and a wet *June*,
Is commonly said to bring all things in tune.</blockquote>

When black ears appear off wheat, it is a sign of a good corn year. Wheat shews itself best in blade, for when it comes to shoot, it looks thin. A wet season, in blooming time, breeds and brings the green fly that sucks the bloom, and causes the ear to miss kerning: but dry weather and a brisk wind, keep the fly off that they cannot settle. But dry weather and a calm time suit the bloom best, and cause a plentiful crop of wheat.

In 1739, a forward sown crop of wheat had done blooming by *Midsummer*, having begun about a fortnight or three weeks before. A little wet does it good, but a great deal damages. In this wet, cold summer, the wheat suffered about this time a week or more small frosts, that caused many ears to miss in their kernels. It was this severe season that tried the strength of most grounds, for those that were poorly dressed had poor crops; and as it happened this summer contrary to most others, the latter sown wheat fared best, as to its blooming, because the forward sown met with a very cold, long, wet season,

son, which stopped some, and washed off the bloom of other wheat-ears, so that thousands of acres of this grain were furnished with only poor, thin, light kernels; but drier weather succeeding, the latter sown had such a good blooming, or kerning season, that it produced a full kernel in a full ear. However, the greatest security of all others against this fatal mischief, is by sowing wheat early, for then the green ears commonly meet with the driest and warmest weather, and consequently the best blooming time; and thus causes both the ear and straw to acquire such a hardness betimes, while the shortest nights last, that neither the dews nor insects can so easily hurt them. In this, and next month, wheat is most liable to these misfortunes.———Dry summers hurt not *England*. That is, it is better by far that the vales feed the hilly country, than that the vales. But there is room for exception, as to the dryness of years, for the winter 1739, and spring 1740, were dry, and so severely cold, that they perished a great deal of wheat, so as to cause it to be sold for eight shillings a bushel, in *May* 1740; and this it did both in the vales, and in the *Chiltern* countries; for, as I remember, the frosts began about *Christmas* day, and lasted a long time, attended with most sharp winds from the north and east, that killed many of the wheat-roots, but most of all those which lay

high

high in ridge-lands: yet that wheat which stood it, enjoyed such a fine dry blooming time, that none, I believe, was ever known to yield better.

The winter, 1733, was so mild, that wheat grew all along very rank and long, till it shot into a small ear, as it always does in this condition: the same, when wheat looks whey-coloured or yellow flagged, as it did all *May*, and till it shot, and after; then it is a true sign it will be small eared: and this last sign being occasioned by wets and cold, as it happened all that *May*, and till about the fourth of *June*, it was so ominous of a bad blooming time, that the farmers would wager before-hand the wheat did not bloom well: and it accordingly happened; for the bloom came out but slow, and in a small quantity, and that was several times washed off, which was a sure token the wheat-ear would not kern well, nor did it; whereas, the year before, the bloom came on the ear as soon almost as the flag burst, and was very thick on it, which brought on a plentiful crop; for unless wheat blooms well, it cannot kern well. In 1741 there was as fine a bloom as ever was seen, from the first to the last, and, accordingly, a most plentiful crop ensued, which caused the market to fall ten shillings in five bushels, in about one month's time, just before harvest.

If wheat looks yellowish in *May*, and continues

nues so throughout the whole month, it will never rightly recover that summer, because then it is on its shoot or earing, and past hopes; but if it is sick in the beginning, and recovers before the end, it may chance to be a good crop; for all wheat should may, or look yellowish in *April*, and be of a black green in *May*: or take it this way; if wheat holds its colour throughout *April*, and *May*, there is no great danger of a good crop. If it thrives in *March*, it generally mays in *April*; if it thrives in *April*, it commonly mays in *May*. When it mays in *April* it is right, but wrong if in *May*; for then it should, towards the latter end of the month, thrive and shoot into ear, instead of maying or yellowing; which is often caused for want of the ground's being in heart, to enable the wheat to withstand the cold and chilly seasons.

But to be more particular; your forward sown wheat often grows so luxuriant in the beginning of winter, as to spend itself too soon, and especially if followed with a mild spring, which causes it to grow so rank as to want strength, in this month, to carry it on, and then, in course, it must have a sickening time to check it, in order to make it shoot with the greater vigour afterwards. Now there are two extremes of weather, that sometimes happen in this month, which prove fatal to crops of wheat; one is frosts, as

it

It happened in *May*, 1734, when hailstones fell successively for three days, which being of a poisonous, cold nature, and very heavy withal, beat down, and hurt the spindle of the wheat, and stalks of pease, making the latter red headed, and causing many to die. This was a season so severe, that on the eighteenth day the isicles were seen to hang at the eaves of houses; but the poorer sort of wheat suffered most: on the contrary, at another time, about the middle of this month, the season having been mild and rainy some time before, the well dressed wheat and barley suffered much, by their luxuriant growth; insomuch that a great deal fell before it was shot into ear, which proved of very ill consequence to the farmer, because vast quantities never did rise again, and did not half kern, especially in the vale grounds, which are the richest soils; and this misfortune proved the greater, where the wheat was youngest and weakest in stalks; for when this happens to wheat at an older growth, it is often strong enough to get up and recover. When wheat is rank too soon, it has tempted many to mow or feed it, in order to bring it under a more regular growth: but this management is better or worse, according to the nature of the grain, land, or weather. But, before I leave this subject, I must observe, that when

Chap. VIII. OF WHEAT.

wheat is sown, at a right time, in a proper soil, and the seasons of the year prove favourable, there are many acres that never may or sicken at all, yet hold their growth and colour from the first to the last, in right order, then such wheat is in its highest perfection. About the middle of this month, 1736, the wheat in the *Chiltern*, high ground, especially, looked short, yellowish, and sick, by means of a very long cold dry season, which ruined many crops that were not well dressed, because the ground had not strength enough to make it run and rally again, as that in heart did, and recovered. There was a field of latter sown wheat, which in the beginning of *March* looked at some distance as if nothing grew in it; yet the ground being clear of weeds, and in tolerable strength, and the wheat not having spent itself in winter too much, in the spring it gathered to that degree, that some of it had fourteen ears from one root, as it grew in a loamy, gravelly clay; and proved an excellent crop at harvest.

A farmer, near *Watford*, had, in the great frost of 1716, a field of wheat sown in two bout lands, and it was allowed by almost every looker-on, that the frosty weather had killed it; and they advised the owner to plough it up, and sow it with barley, saying, if he did not, he would have

have nothing but what grew in the thoroughs that lay out of the wind's power. A plain inftance of the wifdom and goodnefs of God, who, though he made the earth to feem barren for a long time, even in the month of *May*, 1740, yet, in a miraculous manner, kept the weeds down till the wheat recovered a head; at that harveft in many places, there were never better crops, and, even where wheat was very thin on the ground, the ears were wonderful large. I had one that had fixty-eight kernels it it. Some ears were fix and feven fet, and one that I heard of, had eight fet; for it is a rule, that after a fevere winter's long froft, there commonly follow the largeft ears.

CHAP. IX.
OF FEEDING WHEAT.

IT was on account of the great ranknefs of the wheat, that many mowed it in this laft fpring in order to check its further growth: others fed it late with fheep, to keep it down that way; this gave me an opportunity to make obfervations on the feveral managements of different farmers: one mowed it down very near the bottom, in a low, wettifh, black ground, that was a rich foil, and the wheat very rank,

Chap. IX. OF FEEDING WHEAT.

rank, on *Easter Monday*, 1732. the ear being ready to come out of its skin or cover; this backned it very much, and it proved a good crop at harvest: this was done by Mr. —— *Fen*, at the bottom farm, near *Barkhamstead*. The parson of *Auberry* also mowed his about the same time, in an open field, by only cutting off the top-part, and it succeeded.

Another finding his wheat likely to run too much into straw, and too little into ear, mowed it, and found it afterwards to shoot so weak in both, that he suffered by it.

Another says, under such apprehensions, it is best to mow it early enough: that some have mowed wheat three times, and had a good crop; for wheat has two or three knots or joints now if it is mowed late, it must be only just topp'd with the scythe, for if then it be mowed below the bottom knot, its apt to kill it, or at least to cause its new shoots to be so weak, that they will come to little or nothing; but when wheat is mowed high, it shoots from the next knot below the cut, and often comes the stronger; as I have proved in a field of my own, that I did for trial sake. This sort of husbandry is perfectly necessary, when wheat spends itself too much in chalky soils, as it did this summer, 1732, more than in the memory of man; for if it is let alone, it will rot at root, tumble down,

down, and be good for little, as a great deal did; so that at *Risborough* by *Wender*, there was a man had five hundred fifty-three shocks, at ten sheaves each, on five acres of ground, and forced to thresh sixty sheaves for a bushel of wheat. Here then are two extremes in this management; if it is mowed too low when it is high, then it weakens it too much; and if it is not mowed at all, then it is a fault as bad as the other: which made an experienced farmer say, in my hearing, that many spoiled a great deal of wheat this year, by mowing it too low and too late.

Several others eat it down with their sheep, and succeeded well; however, in this affair, there is required a great deal of judgment; for first, if the wheat grows in a rich dry soil, clear of weeds, then it may be made more free with, and eat down from *Christmas* forward. Secondly, if it is a wet, clay, or loamy ground, and the wheat thin, this usage will be hazardous; because the weeds here will be apt to get up and keep down the wheat; nor should it in either case be fed in wet weather; for then the sheep by their tread, will crush it into such hollow, wet ground and spoil it, or else will pull it up with their mouths: the same care is also necessary, that it be not fed too late on such a bottom; for wheat on this cold, wet land, won't recover

so

Chap. IX. OF FEEDING WHEAT.

so soon as that on dryer; and this very case attended my next neighbour, who this last spring fed it so late, that it did not recover itself time enough, but proved smaller than ordinary in the ear; therefore several farmers think it best to turn in betimes, and feed it while it is sweet, which often is from *Christmas* till the middle of *February* or longer, when it begins to be bitter; and by thus turning in betimes, they will eat the weeds before the wheat. I knew a farmer that kept his ewes and lambs so long in, till they eat the wheat up, and left the horse gold weed, but he still kept them on till they eat or spoiled them too; but this so stunted his lambs, that he said, he would never do so again; wethers indeed are far more proper for this purpose, and will, if kept long enough on it, eat up both weeds and wheat, provided the former are not too old; and if this work is done with judgment, it will both thicken the wheat, and make it come up much stronger.

Some thought themselves under a necessity to turn in sheep when the wheat spindles too early; as it will sometimes in *February* or beginning of *March*; but this may be partly seen before-hand, as they did this last *February*, 1731, when many turned in about the middle of the month, and fed it till the fifteenth of *March* following, which did a great deal of service that mild, dry spring;

but as I said before, where the weeds are thick and the wheat thin, this way murders it; because the weeds will then get master, and choak the grain. In chalks and sands this method is not practised at all, for here the sheep will either paw it up, or pull it up with their mouths in such loose, short ground; besides, this soil is not strong enough to carry it so forward as the loams are, so that it seldom or never suffers by being winter-proud in these lean earths.

A gentleman farmer, Mr. *Dean*, living in his own farm at *Nettleden*, that was a dry inclosed *Chiltern* one, ploughed up a meadow; and, on only one ploughing, harrowed in wheat-seed, which came on with such a forward rank head, that tempted him to feed it in the spring season; this gave him a valuable opportunity to enjoy a plentiful feeding crop in a scarce time, and a full reaping crop at harvest, free of any extraordinary charges; which in my humble opinion, where the land is dry enough to admit of thus feeding such a rank crop with sheep, is a far better way than Mr. *Yelverton*'s burning the turf, and mowing and reaping off the top of the wheat in *April* or *May*, to check its too luxuriant growth; and this, because the sheep by their tread and their strong urine, stale, and oiliness of their wool, tread the earth on the roots of the wheat, and fasten it on them so well

Chap. IX. OF FEEDING WHEAT.

as to cause the stalks to stand the faster in their after-growth, keep down the worm, and discourage all other insects from attacking it; and above all, by such feeding in *January*, *February*, *March* and *April*, the rankness of the wheat may be so efficaciously checked, as to be kept from running into any destructive growth, as I am going farther to prove by the following example.

In our *Chiltern* inclosed country of *Hertfordshire*, a farmer gave a field of twelve acres (that had been clover one year before) only one ploughing, and harrowed in wheat seed on the same, without any manner of dressing or manure till *March* following; when he sowed eighty bushels of *London* soot on two acres of the same field, and but twenty bushels on every acre of the rest, as the usual quantity is. But, on these two acres, he sowed a double quantity, in order to feed some fatting sheep on the same late, when turnips and all green-field meat were eaten up; and accordingly this extraordinary dressing answered his intention, by hurdling out the two acres, and feeding his sheep on the wheat till the first of *May*, 1743. And though he fed it so late, yet, by virtue of a rainy season that happened in the latter part of *April*, the wheat recovered, and went on in a flourishing condition. But the next example will shew the great power that a rainy hot summer has on wheat-crops.

The land was common ploughed-ground, which he dunged well for a wheat-crop, and it produced a rank wheat on the ground in the spring, having been before winter-proud, as we call it, which forced the farmer to eat it down with sheep till *May*-day, and then he let it take its chance: yet, for all such late feeding of it quite bare to the ground, a hot wet summer following, the wheat sprang up so fast, and got in such a burthen, that it fell down before harvest, and yielded a poor crop of corn. There is, indeed, this conveniency in sowing wheat on new broken up meadow ground for a first crop, that such wheat, if rightly managed, will produce the best of seed-wheat, because it will be clean from all manner of trumpery; and, though the kernels may be lean-bodied, yet it will sell for an improved price at market, as being the choicest of seed; and therefore Mr. *Yelverton*'s management was very right, as to the getting wheat for a first crop on his new broken up ground; and also in mowing it in *April*, and reaping the top part in *May*: but, for all that, had a wet summer ensued, his crop of wheat would have been of little worth at harvest, as the account observes; but he was fortunate in the enjoyment of a dry summer, else the double dressing (if I may so term it) of the ashes, and the richness of his virgin-earth, would have shewn the difference.

A farmer

Chap. IX. OF FEEDING WHEAT. 213

A farmer dressed his wheat with soot in *November*, in order to eat it off in *January* following. This uncommon action was performed by a farmer who rented a hundred a year, a little more than a mile from my house, which proved him to be a bold, but judicious adventurer, as I shall make appear. The case was thus: this farmer having a large inclosed field of new-sown wheat, whose soil was a chalky loam, that lay flat, and sheltered very much from all winds, was encouraged to sow twenty bushels of soot on each acre of it, in order to bring on an expeditious cover, or head, on the wheat-roots; (for, of all our common manures, none forces so quick, and so strong as soot) that might serve instead of so much grass for feeding and subsisting his flock of sheep in the severe, cold, chilly month of *January*; and it fully answered his end: for, in that month, it had got so large a head of wheat-blades, as fed his sheep for a month or six weeks; which enabled him to continue folding his wethers, all, or most, of the time, on the fed wheat, as it was eaten off, and parted by a row of hurdles. Now, the soil of this field being of a very dry warm nature, the sheep fed on the wheat, and lay in the fold of the same ground, free of that daubing and chilliness, which a clay, or a wet loamy soil, would expose them to, and damage their bodies; and thus did a vast service, in leaving an additional dressing behind them, that
enriched

enriched the crop of wheat, by thickening the land, invigorating its roots, and bringing on a branching growth of ſtrong ſtalks, and large ears.

I am here alſo to remark, that this action was performed with great judgment; for this farmer knew very well, that this piece of dry ground was the cleareſt and freeſt of weeds of any he rented; and therefore he ventured and ſucceeded; for, had this field been of a contrary ſoil, the conſequence might reaſonably be expected to have been the endangering the crop of wheat; becauſe, if ſuch a crop had been obtained in a clay, or wet, loamy ſoil, it would not have done the ſheep half the ſervice it did in this chalky ſoil; and what might have been worſe ſtill, the natural chilneſs of ſuch ground, and the coldneſs of the ſeaſon, might have ſo checked the ſecond growth of the wheat, as to give the weeds an opportunity of getting the dominion of it, and crippling the crop, to the impoveriſhing of the farmer, as I could make appear by ſeveral inſtances, were it proper in this place.

This piece of extraordinary good huſbandry I have been the more particular in relating, as it may remind the judicious farmer, and inſtruct the ignorant one, in the improvement of his dry, huſky, chalky ſoil, which though commonly let for a mean rent, on account of its looſe poor nature,

Chap. IX. OF FEEDING WHEAT. 215

nature, yet by a management adequate to its quality, it may be made to produce as good a crop of wheat, as a vale stiff soil, that is in its original nature far richer than a dry chalky one.

A great farmer fed his double dressed wheat with ewes and lambs. His next neighbour did the same. The first enjoyed a fine crop at harvest, but the other lost his, because the soil of the first was a dry, chalky loam, and richly manured, which caused the wheat, after being eat down, to rally again, and out-run the weed; when the other that was a stiff wettish loam, that had only a single dressing, gave room to the black bennet, and other weeds, to out-grow the wheat to its destruction.

A farmer, whose land joined mine, intending to make the most of his wheat-crop, fed it down early in the spring, and after it had got a new head, fed it a second time bare with his sheep; but the consequence was fatal; for the horse-gold and black bennet weeds came up so thick, as got the start of the wheat; and at this time, indeed, the field appeared green, but with more weeds than corn, to the almost destruction of the crop. Mowing and feeding down wheat in the spring-time, to sustain sheep, or stop its too great luxuriancy, backwards its ripening, and exposes it to the risque of the longer nights.

CHAP. X.

OF THE SMUT.

WHEN wheat is in its green ear, the smutty ears may be discovered as they stand, but more as they are nearer being ripe, and this amongst other sound ears, by their black kernels, on rubbing which, a black powder will fly out and stink. This sickness in wheat happens sometimes to only one side of the ear, when the other parts remain seemingly sound; as was once the case of a whole field of wheat near *Hazlemere* in *Surry*, where only the west side of the ears was smutty, and the rest free throughout the field; which seemingly shews the disease to be occasioned by infectious wind.

Case the second. A man having but one field, it was sown for him with naked wheat-seed, by a neighbouring farmer, who wanting a little more to finish the field, sent for some of his own that was brined and limed. The latter proved smutty, but the former clear, though both were sown in one day. The unbrined might be sound seed, and the brined unsound.

Case the third. A farmer in *Surry* being obliged to cart over one field to come at another, it happened that all the field of wheat proved sound and clear of smut at harvest, except that part

Chap. X. OF THE SMUT.

part which had been carted over, and that yielded a great deal of smutty wheat. This seems to indicate, that a bad tilth occasioned the misfortune.

Case the fourth. A field of turnips, being half eaten off with sheep between *Albollontide* and *Christmas*, was then ploughed up and sown with wheat. The other half that was not eaten off till *Candlemas*, was also ploughed and sown with the very same seed, but neither brined nor limed; the result was, that the first sown proved smutty, and the latter sown clear and free. By the latter seed's being not subject to the severity of a very long, severe, cold winter, and warm weather daily increasing on the same, I am of opinion it prevented the misfortune that occurred to the first sown. A case that happened to many by the violent frosty winter, 1739, who never had any smutty wheat before.

Case the fifth. There was some wheat sown on a dunghill for a trial, and it proved all smutty. It seems a plain reason, that the great heat of the dung cankered the kernels, and occasioned the misfortune.

Case the sixth. A gentleman who keeps no sheep to fold, and sows only so much wheat as just serves his large family, dresses his ground with only his coach-horses dung; and, though he brines and limes, he has smutty crops when his neighbour's are clear. The case is plain.

Case the seventh. One of my neighbours, an ancient curious farmer, not only changed his seed, but brined and limed it well, yet, the year 1740, had a smutty crop. This seems to be owing to a long frosty winter, cold spring, and dry summer.

Case the eighth. Two fields, whose soil is a mixture of white clay, and a hurlucky chalk, which lie on the hanging of a hill, sheltered from the north and east winds, are observed frequently to produce wheat that is smutty. The worm, or want of sufficient air, or the running water off the hill, may canker the roots of the wheat.

Case the ninth. If land is dunged with stable dung just before it is sown with wheat-seed, it is apt to breed smutty wheat. An eminent yeoman, near the village of *Barly* in *Hertfordshire*, and many others that I have met with in my travels, never dung their land the same season they sow it with wheat; if they do, they say they seldom fail of having smutty wheat, though they change, brine, and lime in the common way; therefore they lay on their dung the year before, and plough it in for sowing the same with either oats, barley, or beans. But in the western parts of *Hertfordshire*, and so in the vale of *Aylesbury*, they

Chap. X. OF THE SMUT. 219

they always lay their dung on it for a wheat crop the same summer, yet observe to do it on the first stirree, which is commonly performed in *May* or *June*, if they are good husbandmen.

Case the tenth. A gentleman's bailiff in the *Chiltern* country, in *September* 1740, on a presumptive notion that smutty wheat would produce a sound crop, if it was thoroughly brined and limed, bought a sack that had many smutty kernels in it; but, after trial, he found his expectation crossed, for it returned him a lamentable smutty crop.

Case the eleventh. By a loamy, gravelly, inclosed field of four acres, there grew a spinny wood. This field was sown with wheat, and about half an acre of it that lay next the wood was very smutty, but all the rest clear, though it was sown with the same seed at the same time. This seems to be occasioned by the wood's retaining the fogs, or keeping off the freedom of winds, or by the suction of tree roots, that may impoverish the contiguous ground, and starve the growth of the seed.

Case the twelfth. It has been observed, that when wheat has been sown late, it is not so liable to be smutty as that sown more forward.

Case the thirteenth. A yeoman that lives in the *Chiltern* country had a smutty crop of wheat in the dry fummer, 1740, and when he fowed wheat the next time, his days-man, by miftake, took the wrong fack in the dark of the morning, and fowed the fame fmutty wheat; and it happened that, in the following harveft, he had half his wheat-crop fmutty.

Cafe the fourteenth. A farmer, that lives about a mile and a half from me, declared his farm was never troubled with fmutty wheat for feventy years paft, till 1741, when his crop proved fo fmutty, that he had been at *Hempftead* market four times, and could not fell it. He only fprinkled water on his feed and limed it.

Cafe the fifteenth. A certain farmer took fome feemingly found kernels out of a fmutty ear, and fet them in his garden againft other wheat-kernels that were perfectly found: The firft proved fmutty, and the latter found.

Cafe the fixteenth. My near neighbour had all his crop infected one year with fmutty ears here and there, but he ventured to fow the fame feed after brining, fkimming, and liming, and it proved a found crop. The fkimmings he alfo fowed by themfelves, and they proved all fmutty.

Cafe the feventeenth. I was told by a farmer that he tried the following experiment: He wafhed his fmutty wheat-feed in three feveral waters prefently after one another, till he had wafhed all, or moft of the fmut out of the kernels,

Chap. X. OF THE SMUT.

nels. Then he steeped the seed immediately in brine a few hours, and after he had drawn off the liquor, he limed and sowed his wheat, and had not the least smut the following harvest. This is certainly a much surer way than what I knew

(Case the eighteenth.) A silly obstinate farmer practised, viz. He put his smutty wheat-seed into brine, and after it had stood a night, he drawed off the liquor, limed, and sowed it, and had a smutty crop in return, 1741, because the smut that is soaked, or washed off, tinctures the liquor, and consequently infects the sound seeds that are among the heap, as appears by the black glewy mucilage or substance that remains in the bottom of the brine after soaking such smutty wheat.

Case the nineteenth. Another farmer was of this opinion, that it was those kernels that grow in smutty ears and appear sound, that produce smutty wheat, and not the small sound kernels; for as the straw or stalks of such smutty ears are generally rottenish at harvest, their ears break sooner off than those growing on sounder stalks, and are picked up in great numbers by gleaners; when farmers buy such leased wheat to sow for feed (as is commonly done, because they think among such wheat there is the least seeds of weeds) they run a great risk of having a smutty crop.

crop. Mr. *Tull* is very particular on this matter; his words are thefe:—" Smutty grains will not "grow, for they turn to a black powder; but "when fome of thefe are in a crop, then, to be "fure, many of the reft are infected, and the "difeafe will fhew itfelf in the next generation, "or defcent of it; if the year wherein it is "planted prove a wet one." Page 250.—The wheat that grows on a two years clover lay, feldom or never produces a fmutty crop.

The conclufion of this chapter, fhewing, in fhort, what may occafion a fmutty crop of wheat.

Firft. It may be occafioned by the weaknefs of brine; that ought to be ftrong enough to bear an egg.

Secondly. By the weaknefs of the lime, which is beyond the brine, for fecuring a wheat-crop from fmut.

Thirdly. By fowing one fort of feed in the fame foil too often.

Fourthly. By a very frofty winter, a very cold fpring, a very wet, or a very dry fummer, or by infectious winds.

Fifthly. By wheat growing very thin among many weeds.

Sixthly. By a rough, four, bad tilth.

Seventhly. By infected feed that grows in the fmutty ear, and yet appears to the eye found and clear,

Chap. X. OF THE SMUT. 223

clear, or by sowing pepper-wheat, or that damaged by insects, or burnt in the mow, or of too great an age, or too small underling wheat-seed.

Eighthly. By the heat of dung that lies along with the seed in the ground.

Ninthly. By the use of stale or urine in the preparation of the wheat-seed.

Tenthly. By a small red worm that is very apt to gnaw the kernel or blade of young wheat, and thereby causes the ear to be smutty, or kills it intirely, as I have known it to do in a chalky gravelly soil, in the years 1740 and 1741, about the month of *November*, before the frosts came on.

Lastly. In very low vallies, great floods, and their continuance, often corrupt the roots of wheat, and cause smuttiness, as well as great rains do that fall about the blooming and kerning seasons, so that the ears, as well as the roots, of this golden grain, may be damaged by too much wet weather, and brought into a smutty condition; for, undoubtedly, all those causes that hurt either roots, stalks, or ears, in their green growth, tend towards infecting this corn with that stinking black sickness called Smuttiness. After the great frost of 1739, we have had the forward parts of three summers very dry, and the latter parts very wet, so that for the three harvests, 1740, 1741, and 1742, there has been more smutty wheat than ever was known in the memory of man.

Burnt wheat is next to smut in its nature, and would have been such had the seed been more damaged, or the cause been more imperfect. It grows in a bunchy short ear, that contains oftentimes some of this pepper-wheat kernels, and some very sound ones; and as they grow in one and the same ear, I am apt to believe it is either for want of a sufficient nourishment at root to perfect the whole grains, that some of them thus prove defective, and grow into smutty wheat; or it may happen by blights that take that side of the ear wherein they grow. However, it is certainly of very ill consequence, when a crop of wheat has too great a quantity of those black pepper-wheat corns in it, because they make a sack of wheat look pye-bald, help to give a brown cast to the meal, and therefore is oftentimes rejected by the buyers. Upon these accounts it is, that a nice farmer will never sow leased or gleaned wheat, for it has been proved by several that have made use of it, they never are free from a great return of the same sort, if they sow such gleaned wheat-seed, by reason there always are many pepper-wheat corns, or small imperfect underling sound ones in their ears, notwithstanding all the liming and brining that are applied for preventing the misfortune. Now, whether these pepper-wheat seeds ever grow again, is a question with many farmers. I confess, fefs, I never gave myfelf the trouble of nicely trying it; but as there is a little fort of flour in them, they may poffibly grow, yet it is impoffible they fhould produce found kernels, becaufe there is not farinous fubftance fufficient to bring them to maturity. Some therefore fay it is thefe that bring forth fmutty wheat ears, but many other farmers think they never grow at all; however, there are few crops of wheat that has not fome of them.

CHAP. XI.
OF HARVESTING WHEAT.

IN a very hot fummer, about twenty-four years fince, the harveft begun fo early, that many had all their grain in by *Lammas-day*; which fo provoked one farmer, that he would get in all his thetches by that time to fave his credit, left he fhould be thought the moft negligent one in the parifh: but he paid dear for this punctilio, for the thetches were fo damp, that he was obliged to carry them out of his barn into the field again, to be dried and houfed in a better manner. But, in the year 1740, it was about the tenth day of *Auguft* before I, and moft others, began reaping our wheat, which according to my obfervation, was the lateft I ever knew; occafioned by a moft fevere winter, and long, cold, and dry fpring and fummer. The wheat that is commonly ripe fooneft with us, is the white, or *Holland* fort. The lateft are the bearded wheats. The red pirks, or red *Lammas*, come in between; but this is governed in a great degree by the time of fowing, the foil, the afpect, and their ftanding nearer or farther from the fouth. When wheat is ftruck or damaged by infects, and the ftraw becomes fpotted and hardened too foon, fo that the ufual afcent of fap or nourifhment is checked and ftopped, the corns rather decreafe than increafe in their growth; or, when the kernels are by mildews glewed, or, as it were, bound in their hofe, or fheath, fo tight, that they cannot enlarge and grow bigger; I fay, then we commonly reap fuch wheat very forward.

In the next place, when wheat is fallen down, efpecially when this happens while the ftalk is green, as it often does by the violence of winds, the continuance of great rains, or by the largenefs of the crop, then the fap cannot afcend to feed the ear, and is enough to oblige us to reap it the firft of any, left fuch laid wheat grow as it lies; I mean, left the kernels fprout in the ear, by wets or the dampnefs of the ground, that in this pofture they lie very near to.

When

Chap. XI. WHEAT.

When wheat is much ftruck or mildewed, fome reap it as foon as it is full kerned; for, as I faid, the longer it ftands, the fmaller it gets, as has been proved by letting a piece of it ftand, after the reft was reaped. The ripenefs of wheat is eafily known to the meaneft ruftic, by the whitenefs, brownnefs, or rednefs of the particular forts of ftraws and ears, and by rubbing out the kernels of an ear in one's hand. However, none ought to be reaped till the milk is out, and the corns be hard.

Reaping wheat early, gives the pirks and *Lammas* fort a bright golden colour, which is fo agreeable to a wheat buyer, that he will give more for fuch, than if it ftood till full ripe, becaufe it will weigh heavier, and yield better flour. But when it ftands too long, it becomes a greyifh red, and its flour will be deadifh, unpleafant, and lighter. A farmer reaped it almoft greenifh, and fo early, that many told him he would fuffer by it; but he faid he never had finer coloured wheat, nor any that fold better at market. The fame I experienced laft year, 1741, when my chief reaper told me it was not ripe enough to cut; but, as it happened, there was not a brighter finer fack of wheat brought to *Hempftead*. Indeed, had it been reaped greenifh, the kernel would be apt to fhrink, be guttery, and more
muft

must go to fill the bushel; yet, if this is not done in too great an extreme, it is better so than when it stands till it is too ripe, for then the kernels will lose their bright colour, get a thick skin and blackish ends, and be very apt to shed at reaping, binding, and carrying; and, if wheat is to be sown on the same spot of ground for a successive crop, such shed kernels, very probably, will beget smutty ones. Accordingly, it is our general method in *Hertfordshire*, to begin cutting before the wheat is full ripe, not only for the foregoing reasons, but also for enjoying the weather in its longest days; and where a farmer has great quantities of this grain, that he may get to the end of his work in due time. Others make it their rule to begin reaping if the wheat is clear of weeds and straw dried; then they reckon it fit to bind as soon as reaped. Others venture to reap when they think the wheat is ready, though the weather is discouraging, because, when it is cut down, there is a greater opportunity of getting it into barn than when it is standing.

Previous to the moving wheat-sheaves to the barns, a mowstead should be prepared to lay them on, for preserving them from the damp of the earth, and in some measure from the power and mischief of rats and mice; for which purpose,

pose, nothing exceeds a foundation of furzen or whin faggots, with a thin layer of straw over them, because these are so prickly, that it is impossible vermin should make any lodgment therein, or without great pain walk over them; which consequently secures the bottom sheaves that lie in most danger, from their destruction; and the better to do this, we place our first layer of sheaves almost upright, and very close to one another; on the ears of these, a second layer or row must be placed a little sloping, and something short of one another, with their straw ends outermost, and so on, with this caution, that the whole body, or mow, of wheat sheaves, lie eighteen inches, or two feet short of the barn-boards or wall, in order to give sufficient room for a man or cats to go round the same at pleasure.

Thus, by laying the first row of sheaves upright, and the rest in a sloping posture close together, with their tails outward, the whole mow will lie tight and compact, and give the air an opportunity to get in, and very much dry away the dampness that may arise from wheat-sheaves being got in too soon, or lying in too close a body. But where furze or whins cannot be had, thorn faggots and straw laid over them, or fern, or straw alone, laid as a bed about two feet thick, must supply the place of furze.

However, in cafe the weather continues fo long wet that you are obliged to get in your wheat-sheaves damp, make a hole in the middle of each bay or mow, by letting an empty hogfhead, barrel, or kilderkin, or four fquare boards tacked together, remain in the middle of it, till the corn is up to its top, then put it up, and in this manner leave a hollow place or dry well, where, if a moufe falls into it while the fheaves are fweating, it will fuffocate it.

Thus corn cut unripe, or inned not thorough dry, will be delivered from the mifchief which generally attends dampnefs, and caufes fometimes a mow to be on fire, or breeds mouldinefs, or rots both grain and ftraw, or at leaft gives the wheat an ill fcent and coldnefs. In either of thefe cafes you muft expect the lefs price, for the firft thing a wheat buyer does, is to thruft his hand into a fack, and if it feels cold or damp he refufes, and goes to another, or elfe leaves a difcouraging offer behind him.

A ftack, or rick of corn or hay, fhould have a foundation of furze, thorn, or other faggots; the lower and moifter the ground, the higher it fhould be raifed, even from one to three feet; for, if the bottom of a ftack was to lie very near a watery earth, two or three feet of its lower part may be fpoiled. A ftack or rick is laid in form of a long fquare, with its top in fhape of an old-fafhioned

Chap. XI. WHEAT.

fashioned house's roof, for the water to fall quickly off. In this shape wheat-sheaves in the beginning should be laid with their straw ends outward all the way up, to let in the air to the ears of corn, and keep wets and vermin from entering. And the better to prevent rains hurting our stacks of corn, we commonly lay pease, or beans, or only straw, on the top-ridge part of it, and then timely thatch with straw over all; for, in this case, barley, oats, or thetches, are not proper, because either of these will stain and damage the wheat kernels, that is, so mix among them, that they cannot well be got out.

And in this form it is that we stack wheat-sheaves, barley, oats, pease, beans, thetches, clover, sainfoine, and natural hay. But to keep any of these corns the more secure from accidents, some lay them on a frame of joists with fixed boards over them, supported by stone, brick, or oaken pillars, of two feet, or more, high, with square caps of stone or wood upon each, to hinder the ascent of rats, mice, and other vermin, and prevent the mischief of damps and vapours of the earth. But if the pillars are made of oak, then we nail pieces of tin about their middle part, to hinder the claws of vermin's getting up. Others will lay their corn in a long square stack, placed on a frame of wood, erected so high, that carts and waggons may stand under it, and so make it serve for both uses.

Of the management of Wheat after reaping.

As wheat is cut, its reaps are laid even one by another in flat rows, for the sun to harden, and the air to plump the kernel in the ears, dry up the sap in weeds, and stiffen the straw: here then depends a great deal of the farmer's profit; for if he enjoys a kind time in, and after reaping, till he inns his corn, it will fetch considerably more than if it was washed by rains, to a degree of its growing in the ear; yet one or more moderate showers have often proved an improvement to the grain, by enlarging its body, meliorating its flour, and causing the ears to part with their kernels and their chaff much easier than when no rain happens. However, to provide against the worst, all judicious farmers will keep hands sufficient in readiness, that in case too much rain falls, they may be employed in time, to raise the heads of the wheat a little from the ground, by laying them as hollow as possible for the air to have room for entering between and drying them sooner: such management oftentimes puts a stop to the begun growth or sprouting of the wet kernels. But we never wholly turn the ears unless there be great necessity, because, by such turning, the golden colour of the wheat will be much

much diminished, and a washed, pale, dead one lodged in its room. However, after one, two, or three days or more, letting the wheat thus lie abroad in rows or swarths, in a fine day, after the dews are dispersed and gone, we bind up the wheat in sheaves, and then shock them in ten, twelve, fourteen, or fifteen to the shock, to stand in a double row, made by placing one sheaf against another, to farther harden the kernels, and dry out all humidity, that otherwise might remain in the ears, straw, and weeds; but in case there be danger of rain, then many will take a sheaf at each end of a shock, and spread them over the tops of the rest. Others will never cap at all, saying, if it rains and wets at one time, it will dry another; and when all the ears stand up in the air, the danger is not so great but that many will venture it. In *Middlesex* and *Kent*, they generally bind as they reap, in order to preserve that fine colour, which the ears, by lying on the ground, often lose, and also because they reckon it cheaper; for when they hire reaping by the acre, as most do in these parts, it is done for six or seven shillings each; but if wheat lies first some time on the ground (as is constantly done in *Hertfordshire*) before it is bound up, the trouble and charge will be the more; for the *Hertfordshire* farmer refuses to reap and bind it presently, lest the moisture in the corn, straw,

and

and weeds, be bound up in the sheaves; and accordingly, this sometimes happens, so that the middle part of such sheaves will be yellow and hoary, and give an ill scent to the rest of the grain. However, this is a general rule with all good farmers, that to let wheat stand long enough abroad to have its due cure, is the best way, for the following reasons; viz. first, by standing so abroad, the kernel gets plump, and therefore will fill the bushel the better; and, though then it becomes thorough dry by keeping, yet there will be a little swell and improvement both in the skin and flour. Secondly, If the corns had a bright colour by being early reaped in good weather, by standing dry some time abroad, it will improve it, and cause it to fetch the greater price. Thirdly, All such wheat as has had its due cure in the field, will certainly thrash easier, part with its chaff quicker, and grind better, than that inned too soon: and when it is thus prepared by a swell, a good colour, and made easy to thrash, handle well in a sack, and grind freely, then it may be relied on to sell for the most money at market. But when a long rainy season happens in harvest-time, notwithstanding the sheaves stand erect in shocks, and even when they are capped, the whole sheaf may be wet through and the ears grown: in this case we can only unbind and spread them on the ground again

for

for drying by finer weather, till they can be bound up a second time. However, as the weather may continue so long rainy, that wheat cannot be got in dry, there are ways and means to supply the defect in a great measure by the following methods. But, before I farther proceed, I have to observe here, that in some places, when wheat is in danger of being spoiled by the rains, they lay it up in one or more parcels in the field, and thatching or otherwise, in order to bring it out and expose it afterwards when the weather is settled for fair: and, thus it will pass through a degree of sweating, if time enough is allowed it, to its great advantage; for such sweating abroad will prevent its much sweating in the barn, which is the worst place to sweat it, because its close lying here (if the wheat is dampish) sometimes causes it to be musty and bad coloured. Make bands to bind the sheaves in a dewy morning, they will not break near so soon as if made in a sun-shiny dry time.

Of curing damp or wettish Wheat.

When great and long rains happen at reaping-time, and when wheat lies on the ground, or stands up in sheaves, the grain commonly becomes very much damaged by the loss of its colour, getting a thick skin by the flour's sticking and being glewed, as it were, to it, it, by repeated wets, which sometimes cause the kernels to sprout, and retain an ill scent, to the great hindrance of its making good bread, or pudding, &c. and by obliging the farmer to take a poor price for it at market. Now, though it is a maxim, that corn had better be spoiled in the field than in a barn, because there is the more chance in the former for its recovery than in the latter; yet, by the several ways I am going to mention, it may be saved from spoiling in either of them; to which purpose I shall begin with what a farmer at *Kenſworth* did in such a case.— *First*, He spread his wet sheaves of wheat over the hair-cloth of a malt-kiln, and leisurely dried them in parcels, till he had his whole quantity cured to his desire; and it so well answered his end, that he sold the same for two or three shillings in a sack more than his neighbours did theirs, that did not take this method; for the corns did not suffer by the little fire that was made for them, because the chaff in the ears defended them from the smoke and too much heat. *Secondly*, or a second way is, as it is practised about *Hertford* town and many other places, to cut off all the wet ears, and give them a sun-heat on a hair-cloth, over a wire malt-kiln, which is best done by laying on the ears of wheat as soon as the malt is off, and the fire is extinguished; for the remaining heat will dry them regular and sweet.

A third

A third way, is to dry wet ears of wheat better on a cockle oaft-kiln; this kiln is ufed for drying malt, hops, or wet ears of wheat, &c. in the fweeteft manner; and, for thefe reafons, it gets more and more into ufe with thofe that value the pureft commodities before the worft forts, and is rejected by few or none, except it be for drying too flow for their mercenary profit; that is, it does not by its violent fudden heat blow up or extend the malt kernels, till their fkins are ready to burft, and fo fill the bufhel with fewer of them than if they were dried by a regular gradual heat, which this excellent kiln will do by the fewel of fea-coals confined and burnt in a cheft or trunk of four broad caft irons, an inch or more thick, whofe fmoke is made to pafs about a fpacious room through flues or chimneys of brick built along the infide walls of fuch a room, near the kiln: by which means the air of the place is fo heated, as to dry the malt, hops, or wet ears of wheat, &c. by as gentle a fire as can be defired, even to fo moderate a heat as that of the fun. I have feen more of thefe in *Kent* than any where elfe. Here, if the wheat ears have their due and timely turnings, they will meet with fo fweet and dry a cure, that they will keep a long time in good order. But, if naked wheat kernels were to be dried on common malt-kilns, you may expect them to be fmoaked or tainted

by

by the fire of the place, without a great deal of care, that they will hardly fell in a market. A fourth way is, to cut off all the wet ears, and spread as many of them on a barn-cloth as it will hold at a time in a thin manner, for the sun and air to dry them abroad; and, when one parcel is done, another may be lain out, and so on. A fifth way is, in case the weather prove rainy, to spread the ears very thin all over the barn-floor, and set open every door and the gates belonging to the same, that the sun and air may have a full freedom to pass through and dry them. A sixth way is, if you cannot inn the sheaves of wheat dry, but are obliged to mow them wet, put some dry straw between the layers, and it will drink up the moisture so as to prevent a great deal of mischief. How serviceable then the knowledge of these ways may prove to many farmers, and to the nation in general, I leave to my reader's consideration, who, if he is one that sows a quantity of wheat, has no reason to grudge the price of my book, if it was only for this information; because a man may be so catched by rains, as to have almost all his wheat spoiled, and then out of what must he pay his rent and get a livelihood, since one year's crop of wheat is that which should pay it for two years, and which, by some of these means, may be saved in a great degree from that damage or ruin

that

that might otherwise attend it; being what no other author, in such variety, has ever discovered before.

If you reap wheat when the kernel is soft, let it remain in the field long enough to harden, and have a care you do not inn many weeds with it, for these will keep the wheat damp and make it stink. In case your wheat is wettish or damp when you are obliged to bind it, do not cap it, for, if you do, it will cause the ears to grow as it stands in shocks; but when it is dry and you fear rain, then you may safely do it. It is the firm opinion of a great farmer, that no sheaves of wheat should be capped, unless they stood in *Welch* shocks of twelve in all; that is, in two rows against each other, five in a row, and capped by two sheaves on their top; for then the rows are short enough to be covered.

The Management of getting in Wheat in Vale Lands.

In the northern parts, they reap their wheat with the hacked sickle, as is done in *Hertfordshire*, contrary to what they do in many other countries, where they prefer the use of the smooth-edged reap-hook. In *Hertfordshire*, our women are above handling this tool, and have been so for these forty years past, as thinking it too slavish an instrument for their sex. But in the northern parts, and in many others in *England*, the women

women reap sometimes in an intire company of themselves; but, for the most part, they work among the harvest-men, and so become a mixed company; because a woman cannot so well bind up a sheaf as a man. Thus women make themselves help-mates indeed, and of very great service to farmers, by helping them to inn the fruits of the earth in due season: and so much does custom naturalise women to this labour, here, that even the servant maids in general, as well as farmers wives and daughters, willingly engage in this sort of field-work, and find, by yearly experience, that nothing adds better to their health, than this, and other bodily fatigue, which they are constantly brought up in the exercise of, and inured to in the farming business. It is therefore observable, that, in this country, the inferior sort of women, for the most part, enjoy a fresh-coloured countenance, and a strong, large body, constantly eating no other bread than that made with flour and the bran in it, as it comes ground very small from the mill, or oat-cakes, instead of it; which last sort a little further north, they eat altogether, as having little or no wheat growing near them. And it is here, they say, in jest, their women never die; as much as to say, they live to exceeding great ages, by eating no other sort of bread than oat-cakes, because they are very light of digestion, and clog not the stomach, like the fine wheaten bread.

Here

Here also they are of opinion, that their oatenbread, or cake, is a great antidote againſt the ſcurvy, and accordingly impute the enjoyment of their health to a good old age, very much to its ſalubrious qualities. But to return from this digreſſion, to the ſubject in hand; I have further to add, that, in this country, they reap and bind as they go, by every reaper's making his own ſtraw-band, and tying up the wheat in ſheaves at the ſame time, which they immediately ſhock and cap afterwards, as they do in *Kent*, and many other places. Now ſome I know will wonder, why the wheat is ſhocked and capped ſo ſoon, in fine, dry weather. In anſwer to this, they give this reaſon for it: That, by ſetting up eight ſheaves and capping with two beſides, ten in all, or ſometimes twelve in all, the whole ſhock is better preſerved from the pernicious power of miſts and rains, than thoſe larger ſhocks of fifteen ſheaves can be, as the common mode is in *Hertfordſhire*, which I think a very wrong one. It is alſo in this our ſouthern county, that we ſeldom or never cap as we ſhock, nor afterwards unleſs we foreſee the danger of rains; and to prove that we are wrong in *Hertfordſhire*, and they right in the north, let it be conſidered that, after a ſhock of fifteen ſheaves is made and ſtands ſome days without capping, if it is capped afterwards ſuch late capping cannot ſettle and ſit so close, as that capping does, which is put on as soon as the shocks are set up. But, for further explaining and illustrating the great benefit of this north country way of reaping, binding and shocking at the same time, I shall make the following observations, viz.

First, By such hasty binding and shocking, the wheat is preserved in a great degree from the damage of damps arising from the ground, where if it lies some days in reaps, as the *Hertfordshire* way is, it generally causes the wheat to lose its colour in a great degree; and, if a wet season attends it, then it is apt to grow and sprout to the farmer's great disadvantage, for such sprouted kernels seldom become thoroughly dry afterwards, because the sprouts retain such a moisture as will not dry out, unless by a great force of the sun's heat, and wind, which the wheat is very much deprived of, by lying in inclosed fields: for here damps and wets remain so long as to occasion the wheat to sprout in a very little time, and then it is as difficult to dry it again; I may say, much more difficult to dry it again, than when it lies or stands in open common fields. Now here may arise an objection, that if wheat is bound up in a dewy morning as soon as reaped, it will in course lodge a great dampness in the sheaves. To clear this objection, I have to offer, that though dews often-

times

times fall in great quantities, and wet much, yet, as they are for the most part only a superficial moisture, they are soon dried away, by the heat that such early and close shocking and capping will produce. Their way therefore here, is, not to begin reaping so soon in a morning as we do in *Hertfordshire*, but they hold working later at night than we do, because the evening damps are not regarded by them. And, if the dews are very great in the morning, they let the wheat lie spread in bands, till the sun and wind in some measure dry it up; then in the beginning of the heat of the day, they bind up, shock, and cap, as I said before. But the great benefit of this excellent management is further proved by what follows.

Secondly, By such early binding, shocking, and capping, the degree of heat is retained and increased in the sheaves, when they are so bound, shocked and capped, which certainly is a great means to preserve the fine, natural, bright colour of the grain, so much valued by wheat-buyers; who well know, that, if wheat is of such a good colour, the flour of it must consequently be very sweet, dry, and white, and, on that account, fetch more money in a market, than that which has lost these perfections, as that wheat generally does, that lies several days in reaps on the ground.

Thirdly, Again, this early way of reaping, binding, shocking, and capping, very much prevents the pilfering gleaner from stealing wheat in the field; for, while it lies in loose heaps on the ground, it is more than ordinary exposed to the rapine of the field-enemies, who take all opportunities of robbing the farmer, but most of all when wheat lies thus exposed; for, when it lies thus flat on the ground, and if they come in the night-time and take a little from each reap, it is almost impossible to miss it. But when it stands in shocks, the theft can be much better discerned, where they pull out or cut off wheat-ears, for here the gleaners generally carry a pair of scissars by their side, and a bag before them to put in the ears they cut off from the straw, in order to lighten their burthen, for many times they are obliged to go a great distance from home. But sometimes, and too often, they make an ill use of these scissars, by cutting off the ears of the bands privately, that hang within side the shock out of sight, and also from the inside of the sheaves, which cannot presently be easily discovered. And so prone are gleaners, in these parts, to abuse the farmers in robbing them of their corn, while it stands in shocks, that many of them are provoked to let the poor of other parishes come into their fields to glean, by way of revenge: which they would not do, if their own

own poor would but behave themselves honestly. A certain gentleman that held some of his own land in his hands, for furnishing his house with bread, and other necessaries, had several shocks of wheat entirely carried off his land, in one night's time, by thieving gleaners, who are said to commit this piece of villainy for two reasons. One was, because they had a pique against him, on account of his niggardice, as they termed it: the other was, because his shocks stood at a great distance from any house, which gave them an opportunity to carry off their booty, with the more safety. Likewise, by this way of reaping, binding, shocking, and capping, farmers are emboldened to let their wheat stand a week or two, in fair weather, in the field without uncapping, that it may pass through a small fermentation; for, in this northern part, they never uncap, unless great rains indeed force them to it; and then it is, because the two capping sheaves are quite wetted thorough, which to dry again, they spread on the ground, and so perhaps the wetted sheaves of the whole shock. Thus by letting wheat pass through a small fermentation, while it remains closely shocked in the field, it will sweat the less in the mow or stack, and be delivered from the dampness of weeds, and the power of their pernicious heats: for if wheat, by any extreme or unnatural cause, sweats too much in the mow or stack,

stack, it in courſe leſſens the brightneſs of that colour that ought with our greateſt care to be preſerved: this is not an inſignificant item, if duly conſidered, for it is of great moment to the farmer, to the buyer, and to the eater's intereſt, to enjoy wheat, and its bread, in its beſt condition; which cannot be done, unleſs a right management attends the cut down wheat in the field, for, according to the common ſaying, a great deal happens ſometimes between the field and the barn. Wheat, therefore, ought to ſtand a week at leaſt to ferment in ſhocks, to cure both that and weeds that may be among it; then there will be no manner of danger of mow-burning the wheat, as it often happens to be, when inned too ſoon, to the farmer's great damage. I muſt indeed own, where a very ſmall company of men are, who reap their wheat off four-thoroughed lands, or what is called two-bout ſtitches, as many do in the *Chiltern* country of *Hertfordſhire* and other places: there, I ſay, reaping and binding at the ſame time cannot be ſo well done as in vale or other broad-lands, becauſe, in theſe two-bout lands, each man cannot reap more than one at a time, and therefore cannot perform this without a great deal of trouble more than ordinary; accordingly it is ſeldom or never done. Nor can ſuch reaping and binding at the ſame time be rightly done in three-bout lands, becauſe the

reaper

reaper muft go too far to get enough to make a fheaf before he binds it. But, in four-bout lands, it may be done by two people working on the fame land, at one and the fame time, by laying bands on the ridge, and binding up directly, or as they come back, which laft is the quicker and beft way; and when, they here fhock their wheat-fheaves, they take them from off the two outermoft of the three lands, and place the fhock-row upon the middle land, that the cart may have full room to take and carry them off.

The great importance of curing cut down wheat in the field, is eafily known to the meaneft ruftic, becaufe on the goodnefs of this golden grain depends its fale; and if the farmer fails in this, he fails in his principal prop: for, by one year's wheat crop, he muft pay two years rent, that is, the year the wheat grows in, and the third or fallow year, when nothing grows on the fallow land, becaufe it is under a preparation for the next year's wheat-crop. But admit he has a plentiful crop of wheat, and this wheat proves fmutty, or abounds with pepper-wheat, or with the feeds of weeds, it will in courfe fetch a fmall price at market. But when to thefe, or any of thefe misfortunes, another is added of inning a wheat-crop that has growed or fprouted in the field, after being cut down, and fuch wheat has loft its fine bright colour, as is wet or damp, then it will fetch a very low price in comparison of that which is got in clear of these damages. If a buyer takes damp wheat, it must be sold at an exceffive low price. But in case the best wheat sells at so mean a rate as three shillings a bushel, as it did in *September*, 1742, what chance has such a farmer to pay his rent, by the time his month or harvest-men, his taxes, his yearly servants wages, his butcher's bill, and a thousand other incident expences are defrayed; and above all, when he has lost a considerable sum by the bad husbanding of his wheat-crop, in the manner I have been describing. Surely such a farmer is in the high road of breaking? Yet this was partly the case of several in the western part of *Hertfordshire*, and I am afraid in many places else, in the year 1742, who suffered very much by their mismanaging their wheat crop in the field, after it was cut down; for they were so very eager of inning it, that, after it had lain several days on the ground in open reaps, they bound it up, without capping the shock afterwards, and hurried it in damp, though it proved the finest harvest (blessed be God for it) that ever I knew.

Chap. XI. WHEAT. 249

Of horses, carts, and waggons for harvest work.

All these, or some of them, are so necessary, that there is no getting our harvest corn in without them. Where inclosed fields lie about the house, and on a level, the carriage of corn may be performed in the most expeditious and cheapest manner by only carts, as several do in *Hertfordshire*, &c. but where fields lie at a distance, and there be hilly ground in the way, then the waggon is, of all other carriages, the most convenient; because this can be drawn much safer down a hill than a cart, as not being so liable to be overturned; can be better stopped by chaining up a wheel, the thill-horse works in more ease and security, and a greater quantity of corn is brought home at a time, than is commonly done by a tumbrel cart. Of these waggons there are several sorts. About *Sandwich* in *Kent*, they make use of long strong hutch waggons to do all sorts of work: In another part of that country they have a light waggon with very low wheels, made so narrow in the middle, that they can turn in a very little room, and are the safest sort I ever saw for drawing loads down and along the sides of hills. In *Suffolk* and *Norfolk*, in their heavy sandy land, they work the lightest waggon that is, because almost its whole body is made with round sticks. In *Hertfordshire* we travel

with

with a large, close, high sort. But in many parts of the west of *England*, they use neither cart or waggon, because their narrow, rocky, smooth hilly roads, and other grounds, will not admit of their draught, so that they are forced to inn all their corn on horses backs. The next thing I have here to take notice of, is the employing of these; to do which, I shall shew a great deal by a little; and that is, how cheaply a small farmer managed the carrying his corn in harvest. His inclosed field lay on a level about his house, and having only three carts and four horses, he employed one cart in the field for loading it with wheat-sheaves, another driving on, and a third at home emptying. That in the field had one horse in it, that a driving, three, and that cart in the barn was emptying, by its shafts resting on a trussel.

CHAP. XII.
OF THRASHING AND CLEANING WHEAT.

WHEAT, barley, oats, beans, pease, and other grain, are near got as dry, as they will be, in the mow, in the cock, and in the barn; and as the field work is by the beginning of *October*, for the most part, over, and the weather

Chap. XI.　　WHEAT.　　249

Of horses, carts, and waggons for harvest work.

All these, or some of them, are so necessary, that there is no getting our harvest corn in without them. Where inclosed fields lie about the house, and on a level, the carriage of corn may be performed in the most expeditious and cheapest manner by only carts, as several do in *Hertfordshire*, &c. but where fields lie at a distance, and there be hilly ground in the way, then the waggon is, of all other carriages, the most convenient; because this can be drawn much safer down a hill than a cart, as not being so liable to be overturned; can be better stopped by chaining up a wheel, the thill-horse works in more ease and security, and a greater quantity of corn is brought home at a time, than is commonly done by a tumbrel cart. Of these waggons there are several sorts. About *Sandwich* in *Kent*, they make use of long strong hutch waggons to do all sorts of work: In another part of that country they have a light waggon with very low wheels, made so narrow in the middle, that they can turn in a very little room, and are the safest sort I ever saw for drawing loads down and along the sides of hills. In *Suffolk* and *Norfolk*, in their heavy sandy land, they work the lightest waggon that is, because almost its whole body is made with round sticks. In *Hertfordshire* we travel with a large, clofe, high fort. But in many parts of the weft of *England*, they ufe neither cart or waggon, becaufe their narrow, rocky, fmooth hilly roads, and other grounds, will not admit of their draught, fo that they are forced to inn all their corn on horfes backs. The next thing I have here to take notice of, is the employing of thefe; to do which, I fhall fhew a great deal by a little; and that is, how cheaply a fmall farmer managed the carrying his corn in harveft. His inclofed field lay on a level about his houfe, and having only three carts and four horfes, he employed one cart in the field for loading it with wheat-fheaves, another driving on, and a third at home emptying. That in the field had one horfe in it, that a driving, three, and that cart in the barn was emptying, by its fhafts refting on a truffel.

CHAP. XII.

OF THRASHING AND CLEANING WHEAT.

WHEAT, barley, oats, beans, peafe, and other grain, are near got as dry, as they will be, in the mow, in the cock, and in the barn; and as the field work is by the beginning of *October*, for the moft part, over, and the weather

Chap. XII. CLEANING WHEAT.

weather commonly frosty and snowy, the farmer in course, for these, and other reasons, is obliged to employ his hands in thrashing out, and cleaning corn for market; a work that requires a good workman; for though corn is got in dry, yet if the tasker cannot clean and free it of the seeds of weeds, and other trumpery, the master must consequently be a loser. On this account I believe, I may affirm it for a truth, that there have been two or three shillings lost in five bushels of wheat, for want of a skilful diligent workman's cleaning it thoroughly well; and therefore it is that many of our *Hertfordshire* farmers will give a good tasker seven pounds a year, before they will a bad one five; because the want of the corn's being duly cleaned may thus amount to a great sum in one year's time; for when a crop of corn has the seeds of weeds mixed among it, or if it is smutty, or abounds with pepper-wheat, (which is often unavoidable in some soils) then the art of the tasker is to get out all, or most of such seeds of weeds, and clean it from many of its smutty or pepper-wheat kernels. To do which, there are several ways practised in different countries, as I shall by and by give an account of; and the rather, because no author whatsoever has hitherto done it, although it is a most material article in the art of good husbandry, and so necessary, that our finest bread

bread cannot be made in the pureſt condition, unleſs corn is truly freed from the mixture of the many pernicious weeds that often grow among it; ſome of which, as the crow garlic, or wild onion, the melilet, and ſome others, give the wheat ſuch a nauſeous twang and unwholſome quality, that the bread made from it, is very much damaged; and ſo are gruels, puddings, and other things made from thoſe oats, where ſuch filthy ſeeds of weeds have grown amongſt it; and yet the darnel, the burr, the crow-needle, and others, are often ſeen among grain as it ſtands to be ſold in our markets, which are chiefly propagated by the old *Virgilian* promiſ- cuous way of ſowing the ſeeds of corn, that ad- mits not of getting the weeds all clean out of crops of grain as they grow in the field. Hence it was that there was a neceſſity for the invention of the drilling huſbandry, becauſe it gives the farmer an opportunity to extirpate and deſtroy the growth of the ſeeds of weeds amongſt corn, by hoeing the interſpaces and vacancies between its rows or drills, the greateſt part of the ſummer, when the ſap in ſuch weeds is in its ſtrongeſt mo- tion or circulation, and thereby the more eaſily killed. On this account it was, that the drill- plough and horſe-break were contrived, and get more and more into uſe, for preventing the vaſt charge of weeding corn by the weed hook, tread-

ing

Chap. XII. CLEANING WHEAT. 253

ing it down to perform the fame, and then leaving the weeds to make a fecond growth the fame fummer, and doing their mifchief when corn is at that heighth that they cannot be come at; fo that corn thus fown in the random fafhion, is inevitably liable to have the feeds of weeds grow among it, to the farmer's great prejudice, and the more, if his workman tafker does not get them from out of the corn he is to make ready for a market fale; and that he may be the better capacitated for doing it in the greateft perfection, I fhall proceed to give an account of the firft ftage of it.

There are differences in thrafhing corn. In *Middlefex* they differ in this work from *Hertfordfhire*, becaufe, as their country is fituated near *London*, and they make good part of their rent by the fale of their wheat-ftraw, they are very careful to preferve it as reedy or long as they can, in order to keep the ftraw in its original body as it came out of the field; fo that the length of their ftraight ftraws, when bundled up in a trufs, will be about five feet or more: for this purpofe they fpread their unbound wheat-fheaves, and lay them flat on the barn-floor, with one row of ears oppofite to the other; when this is done, the tafker is ready to make ufe of his flail, of which there are three forts; one that is capped with iron at the end of the hand-ftaff, and turns

a fwingel

a swingel by an iron swivel, that makes it a very durable one, but the iron and the wood do not rightly agree together, for the iron is apt to cut and wear away the wood too soon. The second flail is made with a capping of horn, that is generally allowed to be the best of all others, because this is neither too hard nor too soft, and lasts a great while; the third flail is made with a piece of bent ash, and is the commonest sort of capping that is made, as being the cheapest of all others. The flail being ready, the tasker, as I said, throws his wheat-sheaves from off the mow, and lays them flat on the barn-floor, with their heads and ears against one another, so that when two rows are thus completed, he falls to work, and if he is acute at his business, he will bring up the swingel as close to his head as he can, by which, with the more facility, he is enabled to strike downright sharp strokes with great quickness, for beating out the wheat kernels in the most compleat manner; while a bad workman who, through ignorance, may bring up his swingel in a half roundish swing, will be a considerable time longer about his work, and not so well able to get out all the wheat.

Now the *Middlesex* farmer, beyond all others in *England*, stands most in need of an honest, diligent, and skilful workman, for thrashing out his wheat, because here he is obliged to preserve

his

Chap. XII. CLEANING WHEAT.

his ſtraw in its longeſt and ſtiffeſt condition for a *London* market, inſomuch that it muſt appear in a hard, ſtraight, reedy poſture, or elſe he may carry it home again; but this they are not wanting in, eſpecially when a truſs of wheat ſtraw, weighing thirty-ſix pounds, fetches eight or tenpence, as ſometimes it does, at the higheſt price, and at the loweſt, a groat a truſs: And that their wheat-ſtraw may anſwer theſe purpoſes, the taſker dares not make uſe of thoſe ſweeping horizontal ſtrokes, in the time of his thraſhing, as is commonly done by the *Hertfordſhire, Buckinghamſhire*, and many other thraſhers; if he does, he will beat up the ſtraw in puckers, break it, and render it ſo ſoft and weak, as to be refuſed at market; to prevent this, he makes uſe of only his two hands to ſhake the looſe kernels out of the ears and ſtraw, and thraſhes on till he thinks all wants turning, then with the help of the hand-ſtaff and one hand, he turns the flooring, leaving that ſtraw with its ears uppermoſt, that before lay undermoſt, but ſtill preſerves all in the poſture it firſt was put in. Next he thraſhes away again, till he thinks it is enough done, and then with a fork he ſtirs and ſhakes all the ſtraw, to let the remaining wheat-kernels fall out. When the thraſhing is thus finiſhed, the next thing is to bind the ſtraw up in truſſes, which the workman does directly from off the barn-floor,

floor, as the ſtraw lies in length, and with bands made of the ſame, tucks all up in a very tight manner, ready for loading, and carrying it to *London*; which theſe *Middleſex* men do in a more complete way, than all others that I ever ſaw, for theſe bind ſometimes a load and a half, or more, on one of their large carts, which is fifty four truſſes; a number they could not carry, were it not that the roads are mended by turnpikes, and made, for about ten miles round *London*, ſome of the beſt in *England*.

Now here I have to remark, that wheat thus thraſhed in *Middleſex*, is the fouleſt thraſhed of any in *England*, becauſe, as I ſaid, they are obliged to thraſh it as it lies always one way, as not daring to break the ſtraw more than needs muſt, by any croſs ſtrokes of the flail, according to the practice, I believe I may ſay, of all other counties in this nation, as the beſt way of all others to clean the ears of their kernels; for by firſt throwing downright ſtrokes, and then ſweeping or croſs ones, the flail ſtrikes thoſe ears that were covered and defended againſt only downright ſtrokes, as I ſhall more particularly obſerve by and by. However, though the *Middleſex* farmers loſe a good deal of wheat this way in a year, for want of clean thraſhing it out, the poultry kept in the inn-yards of *London*, find ſometimes a pretty deal among the ſtable litters; and the

more,

Chap. XII. CLEANING WHEAT. 257

more, when they take fome of the fhort ftraws with their ears, from off the heap or thrafhed corn, and tuck into the body of the trufs when they find it a little too light. But before I quit this paragraph, I fhall touch on a matter as it happened to one of our *Hertfordfhire* thrafhers, as he was working in a barn in *Middlefex*. This man, among others, went up out of our parts into this at mowing feafon, as is the conftant yearly cuftom of great numbers of men out of *Hertfordfhire* and *Buckinghamfhire*, where fome of them ftay a month or two together, for mowing grafs and making hay; but it fometimes happens, that great rains fall frequently in the time, which obliges the farmer to find out work for fome, to keep them employed, left they leave him, and he want them afterwards: For this purpofe, one of thefe meadow farmers, who ploughed fome grounds every year, fet the man to work to thrafh wheat, and accordingly he began, and proceeded as he did in *Hertfordfhire*, but as foon as the mafter perceived it, he fell a railing at him, and telling him he had better give him fome fhillings a day to ftand ftill, rather than go on as he did: on this the thrafher afking the reafon, the mafter faid, becaufe he would fpoil all his ftraw, by the fweeping crofs ftrokes that he made. Then the man followed his directions, and thrafhed out the wheat by only downright ftrokes strokes of the flail, as the *Middlesex* way is. In short, the *Middlesex* farmer does not regret the tasker's leaving a pretty many kernels among his wheat-straw, because he cannot well do otherwise in this way of thrashing, for if they thrash all the wheat clean out, they must break the straw and spoil its sale.

The *Hertfordshire Chiltern* country, and the *Buckinghamshire* vale country way of thrashing wheat, is one and the same; they only differ in the cleaning of it: in both, it is of high importance to the farmer, to have a long and broad thrashing-floor; because, in the first place, his tasker can lay a greater number of wheat-sheaves on it at a time, than he can on a lesser floor, which will save him a considerable time in preventing him going twice on the wheat mow for the same quantity of sheaves. Secondly, by a large floor, he has liberty to move on backwards and forwards with striking the largest strokes, when and where he thinks fit, and it is by such downright and sweeping blows with the swingel, that the thrasher strikes out some of the corns almost as high as the rafters of the barn, so that by the force of his downright, and his sweeping cross strokes, he has here the greatest opportunity of thrashing out a great deal of wheat in one day, and this in the cleanest manner. Thirdly, by the full length and breadth

of

Chap. XII. CLEANING WHEAT. 259

of a neceſſary barn floor, the thraſher can clean his wheat much better than in a ſhort floor; and this becauſe, by the long throw of the caſting ſhovel, the crow-needle, the ſticky feeds of that ugly weed the bur, the light corns of that worſt of weeds the darnel, and many other feeds of weeds drop ſhort in the long ſpace that is between the thrower and the barn cloth. On this account it is, that I am ſanguine enough to aſſert, that one throw of wheat over a long floor, is better than two over a ſhort floor, as is well known to almoſt all thraſhers; becauſe, by a ſhort throw with the caſting ſhovel, the feeds of weeds will be carried in great numbers with the wheat, and but few fall ſhort, in compariſon of a long throw; ſo that in a ſhort throw, there is hardly half that advantage in the cleaning of wheat, as in a long throw. Fourthly, in a long floor there is room for two men to thraſh, to the farmer's greater advantage; becauſe two men can thraſh out the kernels cleaner than one man, by their blows coming quicker and cloſer on the wheat-ears, which beat out the corns with the greater force, confine them more to the ſpot of ground, and get out that chaff which a ſingle flail cannot ſo well do.

Two flails working at a time, break the wheat more at once going over it the firſt time, than a ſingle flail can do, which is very material in thrashing this grain; because if it can be thoroughly beaten the first time going over it, it requires the less trouble afterwards; but if the first thrashing is a slender one, it requires the more trouble afterwards; and this is often the case of a single flail in a large floor. However, where a floor is of a lesser size, and admits only of one man's thrashing, I say, in such a short floor, one man may do as much, or more, in proportion, than two; because here there is not room for two men to thrash at a time, and therefore one will be apt to hinder the other's working. When one man thrashes wheat, he at last draws off only a single row at a time, which is tedious; but when two thrash together, they draw off a double, or the whole flooring of straw at once, and thus do a great deal in a little time.

In the county of *Middlesex* they generally have extraordinary barn floors; in *Hertfordshire* there are some very good ones; and so in *Buckinghamshire*. But I am of opinion that, for the most part, those in *Kent* excel. And here I should go on in describing the manner of laying barn-floors for their longer duration, and greater conveniency of thrashing out corn, but that is rather too long a subject, and therefore I shall defer it.

We do not differ, in cleaning wheat, from the *Middlesex* farmers, but proceed in all respects as they

Chap. XII. CLEANING WHEAT.

they do, in the following manner, viz. after we have thrashed out our wheat, we with a caving-rake, of five or six wooden teeth, rake off the offal short straws, and other trumpery, and put it into a caving sieve, through which we pass all the kernels that may be among it, and we can get out; and then throw the remaining gross stuff to one place in order to get a heap out of it, and thrash it out at a more leisure opportunity. When the wheat is thus cleared of its offal or caving, the tasker stands at one end of a barn-floor, and with the broad, light, casting-shovel, he throws it to the other end; when all is done, he sweeps off what we call the Corals, or those wheat kernels that would not part with their chaff in thrashing: and then, in case the wheat is not thrown clear enough of the seeds of weeds and chaff, some will throw it a second time to get them all, or near all, out; in order to which, the tasker is to observe as at first, whether the wind blows against his throw, or with it. If it blows with it, it is wrong for his purpose, because it will blow and force the chaff and seeds of weeds, to keep company in some degree with the wheat kernels: in this case, therefore, he must carry the wheat in a bushel, or throw it back with the casting-shovel to the former place, to enjoy the second cleaning throw in the beſt manner; which is a matter of ſuch importance, that though the barn doors or gates are ſhut, and the wind blows into the barn, yet the taſker ought to throw his wheat againſt the wind. Indeed, where a barn is made ſo very cloſe that the wind can have no power, then they do not ſtand much upon this point; but if they have a mind to throw it a ſecond time, they throw it to the contrary end of the floor, without regarding which way the wind is. After every four or five buſhels are thrown, and the corals are ſwept off, the throwing part is finiſhed.

The next thing to be done is to further clean the wheat by the wheat-ridder, which is a round ſplintered ſieve, worked in a round manner by the taſker's two hands, and who, by the art of working this ſieve, will cauſe thoſe corals, ſeeds of weeds, and other trumpery that eſcaped the throwing labour, to gather on the top of the wheat in the ſieve for his throwing them out, to be kept in a particular parcel by themſelves, to be thraſhed hereafter; and this we call Peggings, being compoſed of thoſe corals that are ſwept off the heap of wheat after throwing; and thoſe corals, ſeeds of weeds, and other trumpery the ridder-ſieve thus diſcharges. Then the farmer, when he has got a ſufficient quantity of them, thraſhes them clean, and grinds them for

his

Chap. XII. CLEANING WHEAT.

his family use, or mixes some of these peggings with some better wheat, for his home consumption: and thus he does by his cavings; for as all wheat that arises from the thrashing of his cavings or peggings, has a mixture of the seeds of weeds, or pepper, or smutty wheat, it is not fit to be carried to market with the better sort; no not a little of it, lest such a little spoil the sale of a great quantity of good wheat. But to return to my subject in hand.

After the wheat has been cleaned, by throwing and riddering, the next, third, and last stage of cleaning, is done by screening it; for this purpose the *Hertfordshire*, *Middlesex*, and most other farmers, make use of the tall wire screen, that stands in a leaning posture of itself; and here the tasker does all his work himself, as he did when he throwed the wheat. (But when he riddered it, I should have told you, he made use of a boy to supply his sieve, by putting more wheat into it with a shovel, as the sieve let it out.) Now this screen having an open wooden receiver, or hopper, on the top part, the thrasher with his casting shovel, puts up more wheat from time to time, as the former passes through it, and so on till all are screened or cleaned. Then the next and last thing he does to his wheat, is measuring it, and at the same time putting it up by the bushel into five bushel sacks, which we in *Hertfordshire* call one load, and is carried on a man's back

back into a cart or waggon ready to be drawn to market. This puts me in mind of what I saw in other countries, where their facks held three bufhels only, others four; but in our country they carry it all in five bufhel facks; and at market, thofe men, called Sack-carriers, carry a fack of wheat twenty poles together before they mount a high ladder with it, to fhoot it into a granary; but as this is very hard and dangerous work, we always give them three-halfpence a fack for their pains.

The vale way of thrafhing and cleaning Wheat.

The richeft land this nation affords is commonly found in vales, that returns the biggeft of crops, with the leaft dreffing; and thus makes the farmer amends for a bad road, and the want of plenty of wood; the contraries of which are commonly enjoyed in their greateft latitude by moft *Chiltern* farmers. As then thefe vale farmers enjoy the biggeft crops of wheat in the kingdom, they ft. nd in need of the ftrongeft and artfulleft thrafhers: firft, to get the corn clean out of their long *Lammas* ears and ftraw, and next to clean it in the beft manner poffible; and this they had need do, and every thing elfe they can, to reach the greateft price at market; for fome of thefe came out of *Leicefterfhire* and *Northamptonfhire*, as well as out of *Buckinghamfhire*, and draw fifty or fixty bufhels at a time to *Hempftead-market,*

Chap. XII. CLEANING WHEAT.

market, which lies thirty or forty miles diſtance from ſome of their homes. They therefore employ the ſtouteſt and ſkilfulleſt taſkers they can get, for thraſhing out their pay-rent corn, wheat; and, as they for the moſt part poſſeſs the beſt of oaken planked floors, they take care to have their ſtraw thraſhed thoroughly clean out; for all, that may be left behind by a ſlovenly thraſher, is loſt to the farmer. On this account, it has been known, that a lazy ignorant taſker has left as much wheat behind him after he had done thraſhing it, as his wages come to, for reaſons I am going to give.

The ſort of damage, I am going to write of, accrues for the moſt part to indolent gentlemen, and thoſe farmers who are negligent of inſpecting into their domeſtic affairs, or unſkilful in judging of huſbandry matters when they do. This damage here is committed by two ſorts of perſons, the menial taſker ſervant, and the days-man thraſher. If by the firſt, he does it chiefly for the ſake of making haſte to thraſh out five buſhels of wheat in one day, and leave it in its chaff, in order to ſpend the remainder of the ſame day in the enjoyment of his pleaſure abroad; for, ſays he, if I can perform my taſk in thraſhing out and cleaning five and twenty buſhels of wheat in one week, what is it to my maſter whether I work or play? But here too often happens a

damage

damage to the master; for the thrasher, to dispatch his task, and have the more time to himself, I say, too often thrashes out the head of the wheat, and leaves that which sticks closest and fastest in the chaff and ear behind, because this sort requires the more labour and longest beating to get it out; and thus, as I said before, all such wheat, so left unthrashed, becomes lost to the master. If the damage is done by the second, it is often done altogether out of laziness, because the days-man thrasher has his wages paid him commonly for his work, be it for any thing he is set about, and accordingly must remain in the barn from six o'clock in the morning till six o'clock in the evening in summer time: therefore, in course, he will make his work as easy as he can, and that is to be done in thrashing, by beating out only the biggest corns, that always come out first, and thus leave many of the small corns behind in the ear, because they require more than ordinary labour to get out.

This puts me in mind how a rascally fellow served a gentleman farmer, or *Yeoman*, as the country term is. The gentleman, to prevent (as he thought) a day's-man thrasher's imposing on him, bargained with him to thrash and clean his wheat for three pence a bushel fit for market, the bargain being made during the gentleman's pleasure: this thrasher found out a way to be too cunning

Chap. XII. CLEANING WHEAT.

cunning for him, by taking this advantage; he threw down a flooring of wheat-sheaves off the mow, as usual, and seemingly before the gentleman's face thrashed in a right manner; but, in his absence, he took the opportunity of beating so much corn out of every sheaf, as would come out very easily, and then bound up the same sheaf, and laid it in a part of the mow, where it would appear as if it had never been meddled withal. And thus he proceeded till he got the quantity that was to be sent to market, in hopes that when the next thrasher was employed, they would lay this fraud to his charge, and forget the veteran rogue.

In vales, as in *Chiltern* countries, their way of thrashing wheat, as I have observed, is one and the same; but, in cleaning it for market, they differ very much. In the vale they clean by the wind-fan, as being the cleanest way of all others; and although our *Chiltern* farmers refuse to follow it, for reasons I shall by and by assign, yet our *Chiltern* oatmeal-makers prefer it beyond all; because it blows away their hulls, and offal-stuff, better than any way they can invent; and, for a farther proof of all this, I shall produce undeniable reasons before I quit this chapter.

After the wheat is thrashed and cavined, so that all the gross short straws and other trumpery are taken out of it, and the wheat and chaff only left,

left, the wind-fan is set near the door of the barn, and a bushel or tub with its bottom upwards placed by it, for a man to stand on it, the better and more conveniently to sieve out the wheat before the wind-fan, which is here placed and worked, to blow away all chaff, seeds of weeds, and the lightest kernels from among the capital big-bodied kernels, in order for causing it to fetch the best price at market. Now they compleatly perform this work, with the help of two men and a boy. One man sieves out the corn, and the boy with a shovel supplies the sieve as quick as the sifter can discharge it. The third turns the wind-fan, and by the quick motion of its sails, produces so strong a wind, as clears the wheat in the greatest perfection; and the better, as the fan is placed fit for obtaining the greatest share of wind to blow away the chaff, weeds, and light kernels of the wheat. The head-wheat is sieved and fanned but once; but the leger sort is sieved and fanned twice, to get all in the best order. Thus two men and a boy can clean ten quarters of wheat in one day, with ease, when one *Chiltern* thrasher, with a boy to fill the ridder-sieve, finds it tight work to clean five and twenty bushels of wheat in one day, by throwing it out of its chaff, riddering it, and screening it.

And now I come to prove that those farmers, who

Chap. XII. CLEANING WHEAT.

who clean their wheat by the wind-fan, clean it much eafier, cheaper, and better than the *Chiltern* farmer: eafier, becaufe the wind-fan faves a great deal of labour, which the *Chiltern* farmer employs in throwing and cleaning his corn: cheaper, becaufe two men and a boy will do a confiderable deal more in a day, than the *Chiltern* men can with the fame number of hands: better, becaufe it is publickly known, at *Hempftead*, and other great markets in *Hertfordfhire*, that the vale-wheat comes, for the moft part, much cleaner from foil or trumpery than our *Chiltern*-wheat generally does; and not only cleaner, but larger bodied; becaufe by the vehement force of the wind which the fan raifes, the moft light wheat-kernels are blown away from the main heap; fo that there remain none but thefe of a larger fize, which always meet with a preference at market: wherefore it is plain, that it is not only their occupying better land than our *Chiltern* or *Hill* farmers do, that caufes them to fell the largeft-bodied corn for the greateft price; but their way of cleaning it adds a prodigious advantage to it, as I have here, I think, fufficiently proved. And happy it is for them, that they do thus enjoy the beft of land, and make ufe of the beft of art to clean their grain; for, as many of them draw their wheat twenty, thirty, or more miles an end to our markets, if they did

not

not fell it the fame day they brought it, it would be to their very great damage, by reafon it would oblige them to attend another market for its fale at a great expence. In fhort, the vale-farmer often does as much at once fhifting and fanning, as the *Chiltern* farmer does at twice throwing and riddering; for, when wheat is in a foul condition by the feeds of weeds and other trumpery, the *Chiltern* farmer is frequently obliged to throw it twice, before he can make it fit for the fcreen. Hence the ufual faying took its rife, that once throwing of fome wheat will do it as much fervice, as two throws will do a worfe fort. That is, when wheat is not very foul, once throwing it will clean it, as much as twice throwing will a fouler fort. In a word, the vale farmer is thought to have the advantage of the *Chiltern* farmer, in the cleaning of his wheat; but cuftom carries it fo much in the country, that there are frequently feen, even in one and the fame parifh, two different ways of ploughing and fowing the fame fort of land, in the management of their cattle, and many other things in the farming way; infomuch that one obftinate farmer, by his continuing his obfolete method of proceeding in hufbandry affairs, has broke, and his fucceffor got money, and went forward, as much as the other went backwards

in

in the world; and this by his more rational and improved way of going on.

CHAP. XIII.
OF THE PRODUCT OF WHEAT.

AS to the quantity of wheat that commonly grows on our common field land, I am confined to an estimation of it according to the nature of the ground. If on a true loam, as we have on our hilly land at *Gaddesden*, we generally have four loads on an acre one year with another for a single dressing, with the fold, cart, or hand; nor does our inclosure here answer better, for the one is as good land as the other: this I call a customary, moderate crop (including the tythe) on our tilth ground. On our gravels, below the hills, as much, both in inclosure and field land, that are here also alike, if they are well dressed; for though this is more hungry, and not of so rich a nature as the loamy, yet by its more kerning quality, we have often as good crops as on them; but it must be more assisted with manures. The chalks in the common fields are now so improved, that they also return as much wheat as the marly-vale soils, even from four to eight loads on an acre: but with this difference, that a double dressing on this white ground,

ground, is but equal to a single one on them. Sands likewise have their peculiar properties, as they are richer or poorer, and require dressing, and grain accordingly. In this, soot nor ashes are proper manures, because of its loose body, that will not retain their light one so long as the loams, gravels, and chalks will; and therefore with fold, cart dung and rags, it receives the best improvement for the growth of rye, peas, turnips, thetches, tills, and white-oats; for wheat, barley and black-oats do not answer here so well as in some loams and clays: but the rye in particular agrees with this dry earth, and also on the chalks and gravels, where it commonly returns the farmer four loads on an acre, for sowing two bushels and a half. This is a grain that certainly exhausts the ground to a great degree, as I have known it to do on a loamy gravel, that I could not recover in six years, after a good crop of this rye was got off, by a tenant that rented the land before me, who sowed it on one ploughing, presently after other grain was got off, in *August*, for only his sheep to feed on in the spring; but this grew so well, as encouraged him to let it stand for a crop, and he had a good one.

The greatest crop of wheat I ever knew in our parts, sown in the random way, was at *Dagnal*, in *Bedfordshire*, where a hog-dealer sowed a little gravelly

gravelly field, containing one acre and half a rood of ground, with wheat, and received from it, in one crop, fifty bushels: but the ground did not want for enough of the best dressing, and that was hogs-dung.

CHAP. XIV.
OF THE SAMPLE OF WHEAT.

I Have largely written on the subject of buying and preserving wheat sound; but have not given the most particular account how persons may know what sort of wheat is the properest to buy, and lay up in a cheap time, against a rising market; nor how to buy wheat that is in a right condition to lay up in store; which is an article of the utmost importance to all such as are ignorant of the same, and yet venture to lay out their money in this commodity, to advance it by an increasing price. And although Mr. *Worlidge*, and the most ingenious Mr. *Jethro Tull*, has given some account of keeping wheat by way of magazining it; yet neither they, nor any other author whatsoever, have ever yet published such instructive particulars as can inform an ignorant buyer of wheat, how to prevent his being imposed on. This is a matter of such consequence, that, for want of a plenary instructive

instructive account of it being made public, thousands have committed such gross mistakes, as have proved not only a great loss to themselves, but, in some degree, a damage to the nation in general. For example: When great quantities of wheat are laid up in a bad condition, or, if laid up in a good one, and becoming spoiled afterwards by ill management, the owner suffers in the first place, and the country afterwards, as being thereby deprived of so much good wheat: a loss that might have been easily prevented, had the person been master of a knowledge sufficient to inform him how he might have bought a right sort of wheat in, and kept and preserved it in a sweet sound condition: and therefore, I have thought it more than ordinary necessary to send the account I have written into the world, to prevent these private and public losses; and the rather, because this affair of laying up and magazining wheat, engages not only great and able farmers, but also gentlemen, tradesmen and others, to become, as it were, merchant-adventurers, in the buying and selling of this ready money commodity; a business very enticing on this and other accounts. I know several that carry it on at this time, without so much as hardly appearing in it in person; and they are those that hire a loft or granary in a market

market town, situated so near the market, that sack-carriers or corn-porters convey the bought wheat on their backs to such loft or granary. One gentleman, last year, 1744, and this, bought up, as I have been informed, five sacks of wheat, that were shot and emptied in a granary hired for this purpose, to lodge and keep it against a rising market; and yet, I believe, never was seen in the market to buy any himself, because he employed a skilful trusty man to do it, one whom he reposed a confidence in, as to his knowledge and honesty; but the better to be on a sure footing with his agent, he pays him, as I understand, so much a sack for buying, looking after it while it is in the granary, and selling it afterwards. Another is a lord's gentleman, who, having saved in his service a sufficient quantity of money, employs some of it in this way; and who, being ignorant himself of buying a true sort of wheat to lay up, employs one that gets part of his bread by carrying sacks of corn on a market-day for farmers, to buy a proper sort of wheat to lay up for him at the lowest price, and sell it for him again when he thinks an encouraging opportunity offers. Thus he turns the penny on the lottery, as it were, of buying and selling wheat; which, as it is the king of grain, never wants a market at some price or other, if it is in found sweet order at the time of sale. The next class of buyers and store-keepers of wheat, that I shall take notice of, are farmers servants: these, by their occupation, are proper judges of good and bad wheat; and as some of them have been good husbands enough to save money by their several years service, employ it in this way of buying and laying up of wheat in a market-town, in hopes to make a profit of it; and this he does generally by some friend, whose business is otherwise to attend at market every week; for he himself has not always time nor opportunity to do it: And some of these have been so succesful as to continue this business years together: And if such an one's pocket does not enable him to buy wheat enough to defray the charge of a loft, &c. then I have known a farmer's servant (who, by his diligent service, has obtained leave of his master to go now-and-then to an adjacent market for this purpose; particularly a tasker, who thrashes out his quota of grain in the usual customary limited time every week) engage himself with a tradesman to help him out; the first to find judgment and money, and the last to find attendance; and in this manner have gone on well, not only in buying and selling of wheat, but likewise by doing the same with other corn. The last sort of buyers or traunters of wheat that I shall here mention, are

Chap. XIV.   OF WHEAT.

are mere tradesmen, who, having spare money, and a fancy to employ it in this business, in order to have two strings to their bow, for a living, set up for buyers and sellers of wheat only; and though they themselves are not sufficient masters at present of the secret, yet venture to be dealers in it, by seeing through other mens eyes, and trusting to their better judgment, in hopes to become such proficients themselves in time, by repeated purchase of this grain, as to learn to know a right from a wrong sort, and deal in it without the help of any. Instances of this there are many to my knowledge, some of whom get, while others lose. But whoever he is that endeavours to turn his money in this way to profit, he ought to be a good judge of what he does; because, if ten or fewer sacks of damp wheat are mixed with hundreds of dry wheat, the whole, if kept any time, will certainly be damaged by them; and if kept long, the little bad will infect the great quantity of good, cause the breed of wevils, and perhaps spoil the whole heap. In short, to prevent then the damage that store-wheat may occasion, both to the private person, as well as to the public, I have drawn up the following account, not built on probabilities and uncertain theory, but published from a market-practice of many years experience, and from one of the best markets in *England* for grain, as *Hempsted* is justly allowed to be; for to this market wheats are brought from great distances; and, I believe I may say, from off all sorts of soils, and perhaps in such variety of species, that few markets beside have the like.

When Wheat is best bought to lay up in Store for long Keeping.

This article is of considerable importance to regard, because wheat cannot be laid up too dry for keeping, and it seldom arrives to a true dryness in barn-stack, nor any where-else, till the winter is past; for although it is housed dry, yet it may receive such a dampness afterwards as to make it unfit to be laid up for store, for want of lying long enough in the straw: therefore it cannot be said to be rightly ready for the flail, and threshed out to keep long in a loft or granary, till *March*, when the winds of this month are generally of such a drying nature, as to affect all things above ground little or more. I say, in *March*, or in any other of the subsequent spring and summer months; for upon this depends very much the good or ill success of keeping wheat long; and to know when wheat is truly dry, sweet, and sound, is the chief art of a buyer and storekeeper; and therefore I shall, in the next place, proceed to give such an account as never was done yet, how wheat may be rightly known to be thoroughly dry, sweet, and sound.

I am

Chap. XIV. OF WHEAT.

I am now come to treat on one of the chiefest articles relating to the preservation of naked wheat in a dry, sweet, and sound order a long time; without the knowledge of which, it is in vain for any person to commence wheat-buyer, and attempt to lay up sale-wheat, for keeping it against a rising and profitable market. This is an art that consists in more than four several particulars, viz. in seeing it, handling it, smelling it, biting it, &c. And first, I shall begin with discoursing on the looks or sight of wheat, as it stands to be sold in sacks in the market. The sense of seeing is perfectly necessary to be employed in this business, because a right colour is an indication, in a great degree, of the goodness of this grain. Wheat that has a bright fine brownish cast, whether it be a red *Lammas*, a yellow *Lammas*, or a pirky sort, is valued the more by good judges, for being of this colour; which shews, that such wheat had little or no rain fell on it, nor while it afterwards stood in the field; consequently such wheat must be got into the barn or stack in exceeding good order; and the better still, if it is cut not too soon, nor too late: if it is cut too soon, the colour indeed may be tolerably good, but the kernel will be somewhat shriveled and guttery; and if it is cut too late, it will have a full kernel, but a deadish colour and flour; therefore, when wheat is almost, but not so full ripe, as to begin to shed out of its hose, it is best to reap; for then it will be plump-kernelled, have a delicate bright brownish colour, and furnished with a lively spirituous flour, that will make excellent bread. To obtain these desired ends, fine weather at harvest is a great blessing indeed; for then the sun shines strong, scorches the kernels in the ear, and thus causes them to acquire the fine coloured plump body I have been writing of. By the sight also, smutty, and pepper-wheat may be distinguished from that which is sound; and by the same sense may all seeds of weeds be discovered, as those of melilot, darnel, burs, cockle, crow, needles, &c. which in many grounds grow up with the wheat, and greatly infect and damage it; for these, or any of these, do it little or more harm, and lessen the value of it either in kernel or flour: and for the better discovery of such pernicious smutty, pepper-wheat, and seeds of weeds, a nice buyer of wheat will make his inspection deeper into a sack of it, than just to turn over a handful or two of the top part; for as I sell all my wheat at *Hempsted* market, I have had an opportunity to see more than one trick played with wheat to cheat a buyer; one of which fell to my lot in the year 1744; for, having then an order to buy several sorts of seed-wheat to send to a gentleman, my sack-carrier informed me,

that

that he faw a fine fack of yellow *Lammas-wheat* to be fold; a wheat that he was fure would pleafe me. On this I gave him money, and trufted him to buy it for me, which he did; and, indeed, it appeared at top to be excellent wheat-feed; but fhooting it out of the feller's into a fack of my own, there was feen, in above half the wheat, fuch a foul mixture of feeds of weeds, that I was forced to grind it for my family ufe; and as the owner of the wheat fold it by another hand, I could have no recompence allowed me. It is for this reafon, therefore, that a judicious buyer runs his hand pretty deep into a fale-fack of wheat, to fearch and fee if the wheat is as clean lower, as it is at top, to prevent this fraud of making the top bufhel more clean than the reft, on purpofe to deceive the eye of a buyer; which leads me, in the fecond place, to confider the benefit of handling wheat before it is bought. Now to the fenfe of handling and feeling fale-wheat, is very much owing the knowledge of underftanding its value: this is fo much relied on by the beft judges of wheat, that they will buy none before they feel it; and on this account it is, that as foon as our *Hempftead* market bell rings at twelve o'clock, for giving the farmers notice to begin to fell their wheat, the wheat buyers directly run their hands into the five bufhel facks of this grain, and if they feel it cold

(as

(as their usual term is for dampness) they generally directly leave that sack for feeling into another man's; and so proceed till they feel that wheat which best pleases them; and that is the sort that feels truly dry, has a plump body, and good colour; and when they have met with such, the question is asked, how much a load the farmer will sell it at (for at this *Hempstead* market, we call a five bushel sack of wheat, a load, being the largest usual quantity that sack-carriers, or corn-porters, commonly carry on their backs) and if the price is agreeable, they buy it; some for grinding directly, and others for laying it up in store against a rising market; for in this town there has been sold upwards of —— sacks of wheat on one market day; which gives employment to a considerable number of persons to deal in this grain, that we call mealmen, because they furnish many of the *London* bakers with great quantities of fine wheat flour every week throughout the year, and thereby cause a brisk trade to the several millers about *Hempstead*; and on this account it is, that there are more water-mills erected on the river *Gade*, near this market, than on any other in *England*, in the same distance of ground. Thirdly, the goodness of wheat may be also discovered by the sense of smelling: This may be justly termed a necessary branch of the knowledge

ledge how to buy good wheat fit for grinding, or to lay up for ſtore; for if a ſack of wheat is bought without firſt ſmelling it, a buyer is very liable to ſuffer by it, becauſe ſuch a ſack of wheat may appear to the eye with full plump-bodied kernels, and have a fine colour beſides, yet unfit to be bought for either grinding or laying up in ſtore; to which I add, that a ſack of wheat may appear clear of the ſoil of all ſeed of weeds, and yet be a damaged ſort. Now to make out theſe particulars, I ſhall begin with ſmutty wheat. Smutty wheat, or rather ſmutty balls, that are found in ears among many crops of wheat, yield a moſt offenſive ſtinking, unſufferable ſmell, if rubbed and put to the noſe; and as many whole ears have nothing but ſmut-balls in them, and others part ſmut-balls, and part pepper-wheat in them, theſe, by being bound up in the ſame ſheaves with ſound ears of wheat, are generally thraſhed together, and thereby mixed with the ſound wheat, ſo that there is no ſuch thing as parting one from the other after threſhing, but the ſmutty balls and the pepper-wheat muſt go together in the ſame ſack to market; nor is there any way to prevent ſuch a mixture, but by drawing out the ſmutty ears from amongſt the ſound ears before thraſhing, and this is ſuch a tedious chargeable work, that few attempt it; if they

do,

do, the found wheat is apt to be tainted, and smell of it, and therefore by the smell this is discovered; and when it is, the mealmen generally refuse it, because it gives the flour of the found wheat, a noxious nasty savour, and, in course, would be a prejudice to the baker, by making his bread to smell of the same. Thus smutty balls yield a hateful scent in the barn, in the sack, in the flour, and in the bread. I heard a tasker-servant say, that he could hardly bear to stay in the barn while he was thrashing smutty wheat-ears among found ones, because they made the place stink to an horrid degree. But this is not all that employs the sense of smelling in buying of wheat; there are melilot and crow-garlick weeds that grow up amongst wheat in some grounds, and as these are bound up with the sheaves, and thrashed out together, they give the wheat a most disagreeable smell; but I cannot say this evil is very common amongst this grain, because few grounds are infected with them, and where they are, they grow mostly in crops of *Lent* grain. I have but one hurt to wheat more to take notice of here, and that is the stinking *May* weed, a weed that, I believe, infects and grows in all sorts of earth (except sand) especially where wheat, barley, and oats have been sown in a coarse tilth; for this weed is the commonest weed we have, and is in many places so common, that

in

Chap. XIV. OF WHEAT.

in wet fummers moft reapers greatly fuffer by it, for then it grows fo rampant, as to be almoft as high as the wheat, which gives it the more power to damage men, wheat, and other corn and chaff. To men it is fo venomous, that it poifons their legs through their ftockings, and their hands where it can come at them, and there raifes blifters, which to get prefent eafe of, they fometimes prick them for letting out the water, and fometimes the fkin rubs off, and thence proceeds fuch forenefs, as hinders their reaping till they get better. To wheat, becaufe when this weed grows thick among it, the reapers cannot well help binding fome of it up in the fheaves; and then the tafker is obliged to thrafh it with the wheat, as the pieces of ftalks and feeds are mixed, and remain among it, till it is cleared of them by thrafhing, throwing, and fifting; and although the wheat is by fuch throwing and fifting cleared of the *May* weed ftalks and feeds, yet it is often tainted with a ftinking fmell of it, from lying fo long with the wheat in the barn or ftack, and from being thrafhed with it. I heard that a miller fhould fay, the wheat that he bought of a certain flovenly farmer, was fo infected with the fmell of this ftinking *May* weed, that it tainted the flour of it to his prejudice; but that when he bought it of the farmer, he did not believe it would have had fo bad an effect. For my part, I do

not

not wonder at it, becaufe I have known a great deal of mifchief enfue from the taint of this very naufeous weed; a mifchief that many farmers are fenfible of to their coft, by its caufing their men to lie idle for a cure, for its damaging their grain, fpoiling their chaff, and feeding their ground, to the increafe of its weed. And on the account of fuch its difagreeable fmell it is, that horfes refufe to eat the chaff that is thrafhed from fuch tainted wheat; for if chaff is but little infected by it, they will not eat it with a good ftomach. For thefe feveral reafons, it is a common fight on a market-day, to fee wheat buyers put a handful of the wheat, they are about to buy, to their nofe, to try if they can difcover any ill fcent among it; and if they find any, they either refufe fuch wheat, or buy it at a very low price; and therefore no tainted wheat ought to be bought to lay up for ftore, no, not even if it has but a very little ill fcent belonging to it, left it infects the reft, and hinders it a profitable fale. Fourthly, notwithftanding all I have hitherto written for arming a buyer againft laying out his money on bad wheat, and to buy that which will pay for being kept againft a rifing profitable market, by firft handling it, to prove whether it is dry enough for the purpofe, to fee if there be any foil among it, to fmell if it be any way tainted; yet there remains another thing to be done

Chap. XIV.   OF WHEAT.

done before a perfon ventures upon buying wheat to lay up, and that is this; that all wheat to be bought on this account, fhould be firft proved by biting a little of it, to know whether it will be dry enough to lie one, two, or three, or more years in a granary, without taking any hurt; if it is, then fuch wheat will bite hard and fhort, as if it had been almoft parched; when it is in this condition, has a plump body, and fmells fweet, it may be bought for a complete cured wheat, that will keep found, years together, free of the breed of wevils, and all other damage, efpecially if fuch ftore wheat is made to pafs through the tall wire-fcreen once a year, to take out, and diveft it of all duft that may get among it by long keeping; and thus, fuch a well cured wheat has been kept good feven years together, as I have heard it reported. To this purpofe obferve alfo the following *Items*: Firft, wheat by feeling it only with the hand, may deceive the buyer as to its drynefs; but by biting it the difference is confirmed. Secondly, If wheat is dryed by the fire in facks, it may be difcovered by bite and fmell; for in this cafe it is apt to bite tough, and fmell of the fire. Thirdly, wheat by long keeping in the common way, and doing no more to it than laying it in a granary, lofeth its ftrength and fpirit. Fourthly, yellow pirky wheat, if reckoned to have a thick fkin, is

a tough

a tough wheat, and not so good to lay up as the better, finer, brown pirky wheat is. Fifthly, That, by the nose, wheat may be discovered, whether it is fresh or stale thrashed; if fresh, it will then smell very sweet; for when just out of the straw it is in its sweetest condition, and then fittest to be laid up for store. Sixthly, That, by some, white wheat is thought to be a more tender sort than either a brown, a pirk, a yellow, or a red *Lammas*, and therefore the more unfit for long keeping.

CHAP. XIV.

RYE.

THIS grain is allowed by one and all that I have conversed with, to require a winter and part of two summers to grow in, and therefore they sow it in *August*, on a fine well dressed tilth, if for a crop; but if only to feed sheep in the spring, they often venture it on only one ploughing after wheat or barley, and harrow it in on broad lands; then, if it comes thick, they feed it in the spring, but not too bare, and let it remain for a standing crop. Thus by eating it with sheep, the rye shoots the stronger, and the rather, for the assistance of the dung and stale; but if it proves thin, the farmer then alters

alters his measures, and ploughs it up for peas or oats. The reason for this early sowing, is to get it ahead betime, that it may better endure the severities of the frosts and wets, that otherwise would endanger its dying away, as being not so hardy a grain as wheat: for this so much affects a dry soil, that it utterly refuses wet low grounds; and therefore, the better to secure it against these misfortunes, where any degree of mischiefs are suspected, they sometimes sow it in stitches on fine tilths; or if in grounds that won't admit of this way, they sow it in broadlands that are ridged up, as they do in the vale, in order to carry off the wets more easily. To prove the veracity of this, the men about *Cheddington* and *Aylesbury*, never so much as attempt the sowing of this grain, as well knowing their labour would be fruitless, both in respect of their marly, clayey, black earth, and their low watry situation.

For as it is the nature of rye to grow best in dry, hollow, warm ground; this stiff, cold, aqueous land, is contrary to its homogeneal quality, as much, for ought I know, as it is to sainfoine, that is notoriously known to refuse all stiff, wet, fat earths; because both their roots, I suppose, would rot in the chilly waters that this soil most tenaciously retains, and that, perhaps, by reason of their spungy tender parts, that are more inherent to them than many others; or else, that their roots disagree with the unctuous marly quality of this earth.

About the twenty-sixth of *September*, 1736, I saw rye high enough to cover a hare, in their sandy grounds, between *Thetford* and *Norwich*; and when they sow their grain early in these hot soils, three pecks, or three and a half, are sufficient for one acre, even for a standing crop, because every kernel generally grows in this loose earth; but if it is sown in stiffer soils, then two bushels are but barely enough; for in these there are commonly some buried that never grow, and others that are carried away by birds.

This may be done by any tenant without breaking through the articles of his lease, which, though it generally provides against cross-cropping the ground, yet by this method he may have a crop free of any forfeiture, because it is only a spring feed for sheep, that dresses, not impairs the ground, and is to be fed early enough off, to give the farmer room for a fallow season to succeed the same year, according to his covenants. It is performed thus: As soon as the crop of wheat, barley, beans, or pease, are got off in this month, immediately give the same ground (which must be a sand, chalk, gravel, or a dry loam) one ploughing, and harrow in two bushels of rye seed, twice longways, and once

once acrofs. Thus you may fecure a good food for your ewes, cows, or horfes, to eat the firft time in *March*, and fo on, at times, till *May*, when you may plough the fame up again, and harrow in turnip feed, or prepare the land for a wheat crop. From this management I received confiderable benefit in the cold dry fpring, 1742.

BOOK V.

Of the CULTURE *of* BARLEY.

CHAPTER I.

OF THE SOIL FOR BARLEY.

THIS is a more difficult grain to get plentiful crops of, than wheat, becaufe of its difagreeing with feveral foils and fituations where wheat will thrive; infomuch that it is an ufual faying, that the land which is proper for a wheat crop, is not fo proper for a barley crop; hence it is that moft of the *Suffolk* and *Norfolk* farmers, in particular, who occupy fandy grounds, are obliged to this grain for the greateft part of their profit; and where they have not vent enough for it, for malt-making, they put it to the ufes of feeding turkies, geefe, hogs, &c. which enables them to pay their landlords; and it is in thefe two counties, beyond all others in *England*, that fome fine improvements in hufbandry may be feen, to the infinite profit of both landlords and

and tenants, which have been brought to pafs within thefe fifty years, ever fince they learned the way of fowing and hoeing turnips in their open, common, fandy fields, which has not only proved a preparative to their fucceeding crops of barley, but fuch turnip crops give them a vaft firft profit befides, by feeding their horned beafts with them to a degree of fatting, fo as to fit them in a compleat manner for a *Smithfield* market, where thoufands of them are fold in a year; and by the cooling fat dung, and fertile urine, that their runts, oxen, and cows leave behind them in the land, they fo drefs and prepare their dry, hufky, hungry, warm, fandy grounds, as to caufe them to return more plentiful crops of barley, of late years, than they had formerly.

It is true, that the fandy foils commonly yield a thin bodied barley-corn, and in the reddifh fands a high coloured grain: yet where fuch fands are fully dreffed with dung or manure, the barley kernel may be improved beyond the fize of thofe that grow in poorer grounds. Thus in the fandy loams about *Fulham*, by the help of the *London* dunghill, they often get excellent crops of barley: but ftill their grains feldom arrive to that bulk, as in the more agreeable, gravelly, chalky, and intire dry loams, which leads me to be the more particular in writing of propagating barley crops in the following different soils and situations.

This soil, in one respect, is the most natural soil of all others for the growth of barley; that is, for causing it to grow with the greatest expedition; and, therefore, a farmer that rents a sandy ground, has an opportunity, beyond all others, to sow his barley seed the latest; for in ten weeks time, in such a soil, a crop of barley has been sown and got off; for as this grain, beyond all others of its size, has roots of so tender a nature, that they cannot penetrate into any hard rough earth, a sand, whose particles are for the most part globular, gives the thready weak fibres of the barley roots, an easy opportunity to make their entrance amongst them, and get their living in the most expeditious manner.

On this account it is, that a turnip crop, or a rye crop, that was sown for, and fed off by sheep or oxen even till *May*-day, will, in this loose, warm, dry earth, give the farmer leave to sow the same ground with barley. But then such a barley crop is not in a little danger of being spoiled by droughts, because this is the most burning soil of all others, and in dry times impedes the sprouting of the barley seed to a great degree, and cripples its growth afterwards, by causing the crop to sprout at different times.

That

That barley feed, which lies deepeſt in the earth, will ſprout ſooneſt; that which lies next higher will ſprout next; and that which lies higheſt of all, will ſprout laſt; by which means there will be three degrees of ripeneſs.

April is not ſo late a month as to deprive the farmer of ſowing ſeveral ſorts of field ſeeds. This month gives the farmer, who is maſter of a proper ſoil and incloſure, an opportunity to diſplay his ingenuity, in ſowing therein a proper ſeed; and in order to do this, I ſhall endeavour to ſhew the practice of other farmers, that have acted in this reſpect before them, and by this means make known the dark and bright ſide of their management; for treading the paths that others have made by repeated trials, and found it the ſureſt and moſt ſucceſsful way of proceeding, may ſave, in a great degree, the riſque of a perſon's deceiving himſelf by taking wrong meaſures; happy, therefore, is he who follows the practice of thoſe who have, by repeated trials, found their experiments to anſwer in the moſt ſucceſsful and profitable manner.

A farmer rented an incloſed farm in a *Chiltern* country, where his ſoil was chiefly a clean loam, not too wet nor too dry, for which he paid ten ſhillings an acre a year, for eighty acres, but had no common to keep a flock of ſheep on to fold,

fold, and therefore he kept about forty ewes for breeding only, which difcouraged him from fowing barley, and caufed him to fow only wheat, peafe, and oats. Now this farmer put himfelf to the charge of digging chalk, which he could do in almoft any part of his ground, and laid it on his loams, that it did fome fervice to: but alas! Chalk alone will not do; for though it is a dry, fhort, fweet drefling, yet it wants the moft neceffary quality of all others, and that is the fertile one; for which reafon a judicious farmer adds dung to it in one fhape or other, as knowing that chalk alone is of a poor nature. However, this farmer, by making ufe of no other dreffing than what his ftable, hog-ftye, cow-houfe, and hen-houfe returned him, had not enough to fpare to affift a barley crop, nor would he be at five pounds charge of buying foot, lime, horn-fhavings, woollen rags, or malt-duft, &c. once in a year, to get fix or eight acres of good barley, and thereby loft a profitable opportunity; for without a fine tilth and good drefling befides, it is a folly to fow barley. Now as this farmer lived but twenty-fix miles from *London*, he might (as many do that live further off than he) have variety of manures for this purpofe; for it fometimes happens, that a barley crop pays a farmer more than a wheat crop, becaufe, when wheat,

by

by the extreme inclemency of the winter weather, is hurt and damaged; a barley crop that is sown at a safer time of year, may anfwer far beyond it, as it happened in the year 1740; for had not the barley-crop fucceeded, in that time, better than the wheat crop, there had been a moſt lamentable famine indeed; but as it happened, the plenty of barley helped out with the fcarcity of wheat. From all which I would obferve and infer, that it is a great lofs to let fuch a farm be without a barley crop; not only for the direct profit that it produces beyond oat or pea-crops, but alfo for the benefit of giving the earth a change of feed, which is of vaſt confequence to a farmer, becaufe there is no ground whatfoever but affects this piece of hufbandry: and whoever does not practife it, will be his own enemy. In the vale of *Ayleſbury*, their farmers, fome years ago, thought themfelves abfolutely in the right of it, when they fowed their blue clay, and ſtiff, wettiſh, black loams, with only wheat and horfe-beans; but fince fome of them have fown barley by way of change of feed, they have found their account in it, and been convinced that change of feed is one very great means to come by plentiful crops of corn.

A farmer, like many others in the world, rented a farm in the *Chiltern* country, that lay
about

about twenty-six miles from *London*, whose soil, for the greatest part, was a red, stiff, stony clay, mixed with a loam, so that it might be called a clayey loam, and like the yellow clay, is of the most hungry, cold, and worst sort of clays. In this soil there have been several attempts made to get crops of barley, and to this purpose neither repeated ploughings were wanting to get the ground into a fine tilth, nor manure or dung to fertilise it.

But notwithstanding all requisite preparations were made use of, for obtaining good crops of barley, the farmer was discouraged from making further trials, because of the several fatal accidents that attend barley-seed sown in such a soil. If a long dry time directly succeeds the sowing, then such a clay forms a crusty surface, and hinders the infant barley blades from making their way into the air; and if a wet cold time succeeds, then the chillness of the weather, joining the cold nature of the clay, starves the barley, makes it look reddish instead of a deep green colour, and causes it to pine till great part of it dies away; so that this, and his neighbouring farmers, who occupy the same sort of clayey soil, have been so discouraged from sowing barley, that they have forborne any farther attempt. This engages my pen to repeat what I have formerly made appear, viz.

Gravelly

Gravelly and chalky soils are very well known to be the best of all others for producing a fine, white, thin-skinned barley-corn; and, therefore, when a farmer possesses such an earth, he has a great deal of reason to encourage the growth of barley crops in it; not only for the benefits I have mentioned, but also for that excellent quality more incident to a gravelly earth than any other, which is, its proneness to kerning or corning; for this soil, if it is well dressed, seldom or never fails of producing the largest crops of barley. But, when I mention dressing of this soil, I would be understood, I do not mean a common scanty dressing; no, as this is one of the loosest and most happy sorts of earths, it is absolutely necessary to bestow upon it the greatest assistance, that is, a double dressing, as is often done by several of those able and judicious farmers I know, who rent much of this soil, and, accordingly, receive in return double crops of this grain, when their neighbours, who give it only a common slight dressing, receive but half the quantity at harvest; for, by such double dressing, the seed is so invigorated in its first vegetation, that it presently runs into a branching growth, and covers its root with such a defence against such droughts, as to cause them to produce an early crop.

It is this gravelly soil that yields us the sweetest of turnips, and therefore is mostly sown with turnip seed and barley; for as soon as the turnips are eaten off, they give this earth only one ploughing, and harrow in three bushels and a half of barley-seed; but if the ground is more than ordinarily poor, four bushels is the quantity. Then where a farmer has the excellent conveniency of running his fold over the seed, as soon as sown, and can carry it forward, till the whole field is finished, he will find it the best of dressing for this loose, dry, gravelly soil. But, as few farmers have this opportunity, we commonly, about *Gaddesden*, dress our gravelly barley-ground, with either dung, or *London* coal-soot; but, as often as we can, we do it with the fold; and, when we get a crop of barley, from off this soil in a right order, we seldom ever fail of finding not only a ready market for it, but also a shilling or two in a quarter more than for that barley that grows in loams or stiff earths: and this trial falls to my lot, more or less every year, as being owner of several fields of gravel, chalk, loam and clay, which leads me to give an account of sowing barley in loamy earths.

As this is a stiff earth in comparison of a sand, a chalk, and a gravel (and is, indeed, the stiffest of all others, except clays) it receives a different husbandry from them to sow barley-seed in, for returning crops that may answer the farmer's wishes. For this purpose, therefore, there must be more ploughings performed in this earth, than are necessary to be done in them, in order to bring it into a fine tilth, and make it sufficiently porous to receive and nourish the fine tender capillary roots of the barley; because these roots are of such a soft, thready nature, that they will never thrive, unless they have a full freedom to push them easily into the earth; so that though a farmer should bestow a double and treble dressing on this earth, for making it return a large crop of barley, yet it will be to little purpose, unless the land is brought into a perfect porous fineness, by the plough and harrows.

CHAP. II.

OF TILLAGE FOR BARLEY.

IN loamy soils we sow wheat, barley, oats, pease, beans, artificial grasses, turnip and rape-seed, &c. and after the two latter have been eaten off with sheep, we sow barley-seed, upon one, and sometimes two ploughings: some commend the ploughing the turnip ground once, and others twice, for a crop of barley: if it is

ploughed

that, on one ploughing up of a stiff turnip earth, a good crop of barley may be got, as a neighbour of mine had, who sowed his ground on only one ploughing after turnips, which proved so clotty, that he was forced to chop it afterwards with houghs, and then laid on horse-dung all over the top; about a week after he rotted it, and the following season proved very wet, so washed his top dressing in, and, by its cover, so hollowed the ground, as to return him a very plentiful crop of barley. A *Chiltern* inclosed field of four acres and a half, a field half gravel and half cl..y, had just before the last ploughing a great deal of dung laid on it, that was ploughed in; but, the ground rising almost in whole or flitch thoroughs, buried the dung, and thus was the cause of a very thin crop of barley; which made the farmer say, he would ever after lay his dung on the top after sowing. Now, how obstinately ignorant this farmer was, may appear by his next neighbour's management, who every year, in the same sort of ground, ploughs his dung in for barley, and commonly has very good crops; but then he takes care his ground is very fine and hollow, and his dung very short and rotten, when he lays it on, else the weak roots of barley cannot strike into it, to grow to any perfection. He was obliged to draw, to feed his sheep with them

them on his meadow ground; but this spoiled his barley-crop; because, where he drawed his turnips, the water lodged in the holes, soured his ground, and made it so unfit for barley-seed, that he had but a trifling crop in return.

In the parish of *Studham*, that lies part in *Bedfordshire*, and part in *Hertfordshire*, a great crop of barley was raised, chiefly by good ploughing, dressing, a large crop of turnips, and a kind season of weather. The soil of this field was a gravelly loam, that was extraordinary well dressed with a flock of wether-sheep, to the number of about three hundred; and, when it had been well folded, it was sown with turnip seed, that produced a prodigious large crop, which being fed good part of the winter with fatting sheep, they dressed it a second time with their dung and urine, and thereby got the ground into extraordinary good heart; insomuch that with only one ploughing of it in *March*, and harrowing three bushels and an half of barley seed on an acre, he got nine quarters of barley from off every acre throughout the field, and the straw was so stiff and strong, that at harvest it cut like fern. This account I had from one of this farmer's thrashers, who protested to me, that he and another threshed out twenty bushels of this barley in one day; and he said, that he was sure, if he strove hard, he could thresh twenty bushels of it out in one day himself, it yielded so well.

A very antient man, living in our parish of *Little-Gaddesden*, told me, that in the year 1682, there was no rain till *Midsummer-day*, which caused the barley-seed to lie in the earth till that time, almost as dry as when it was sown; but a great and violent rain falling, fetched it up, and there was a good crop at harvest. And in 1744, it happened to be somewhat the same with barley and pea-crops; for both these sorts of crops were thought by many farmers to be lost by the long duration of dry weather: and yet, when great rains fell in *June*, they brought on prodigious large crops. But then the barley that these two years produced, were of several ripenesses at harvest; because that seed which lay lowest in the earth came up first; that which lay next highest succeeded; and that which lay uppermost was greenish; so that there was abundance of bad malt made, and as bad drink brewed: for without good malt I defy the greatest artist to make good drink: and how much this article concerns the health of human bodies, I leave at present the reader to judge.

Rolling.

In case your land lies rough and clotty, the spiky roller may be of service to go before the harrows,

harrows, for reducing at once the furly glebe into a fine condition; but where this new invented excellent machine is wanting, then ufe the common wooden roller, and your harrows directly after that, and fow your barley; but never roll it as foon as fown, as fome injudicious men have done, but roll it fome time after when it is about half a hand high.

CHAP. III.
OF MANURING FOR BARLEY.

AS barley is a grain that cannot well grow too thick, it requires a more than ordinary affiftance, to fupport its crop, efpecially when it is fown in the old dry way; for, when the ground is not dreffed or manured, there is little hopes of a good crop. Wheat, and moft other grain, will profper with lefs dung or manure than barley will; from hence it is, that many farmers have fallen into the miftake of fowing a great deal of ground with barley-feed, and been at the charge of half dreffing it, in hopes that, with the help of a kind, warm, rainy time, they may enjoy a plentiful crop: but alas! many have deceived themfelves on this account, and been convinced by lamentable experience, that had they fown but five acres where they did ten, and put all the dung on the firſt, that they did on the laſt, they would have had a greater return at harveſt. What a loſs then muſt this be, enough to break a ſlender-pocketed farmer, who ventures to ſow twenty or thirty acres of ground with barley, under a half dreſſing? And eſpecially one that rents a vale-farm, becauſe, if he prepares not the ground, it is great odds, if his barley crop is not ſpoiled by two extremes of weather; one is, that in caſe a very dry ſummer ſucceeds, their ſtiff blueiſh, or black clayey, or marly ſoils will crack and ſtunt the growth of the barley-crop; and, if a very wet, cold one happens, the waters may ſtunt or kill a great deal of barley, which grows near the water-thoroughs; a caſe well known to the coſt and damage, ſometimes, of thouſands of vale farmers, who, by the laſt means, have little elſe to mow beſides that barley which grows on the ridge part of their acre and half acre lands. Now to prevent theſe fatal diſaſters in a great meaſure, well dunging or manuring the land is abſolutely neceſſary for obtaining a full crop of this univerſal grain, by enabling the ſeed to make an expeditious firſt growth, in order to bring on an early head to cover and ſhade the roots againſt the violence and damage of too great heats and droughts, and withſtand the chills of inundations of waters, that frequently happen in the low ſituations

ations of vale-grounds, and ſtagnate the roots of the barley, as many of the hill-farmers are eye-witneſſes of, when they behold vale lands almoſt covered with water, that appear like part of a ſea.

If, as ſoon as your barley is ſown, you fold your ſheep on the ſame, it will cauſe 't to branch, and grow faſter. But as this huſbandry may meet with its inconveniences, as well as be attended with ſucceſs, I ſhall here relate the two following caſes, which may be worthy your obſervation, viz. A farmer by me having had but a poor crop of turnips on a loamy ſoil, that before had been but half ploughed; when the turnips were eaten off, he gave the ground one ploughing and harrowed in his barley, on which he directly folded: But he had no barley in many places, and but a ſmall crop in the better part. The reaſon was, that, the land being four and clung, the barley-ſeed had not room to be buried deep enough by the harrows, ſo that the feet of the ſheep trod it out of the ground inſtead of treading it in; and this misfortune was the greater for the dry weather that followed after; becauſe by this, the ſeed was the eaſier diſplaced by the tread of the ſheep; whereas had it been a wettiſh time, it had not been ſo much raiſed out of the earth.

The other caſe was thus: at the ſame time I had

had a field of turnips, that was a good crop, which, as soon as eaten off, I gave it only one ploughing, and harrowed in my barley-feed, on a loamy soil in *April*, and immediately folded on the same, till the remaining part grew above my shoes in this month; and it proved an excellent crop. The reason was, that my ground was made a fine tilth by several ploughings before the turnip feed was sown, which by houghing and the great cover of the leaves, the earth was kept in such a hollow condition, that the feet of the sheep helped to sink the seed lower, instead of treading it out of the ground. Therefore, none ought to fold sheep on their barley, unless the ground is in a good tilth, and lies loose; and then you will also avoid the common danger of spoiling the barley, by the feet of the sheep, while it is chipping or sprouting, because, when the ground lies thus hollow, there is a sufficient quantity of mould to cover and guard it from such damage.

Likewise observe, that in gravels, and other binding grounds, if it is wet weather when the fold is over the barley, the grain will be so fastened in by the mortarising nature of such soil, and the feet of the sheep, that a great deal will never come out; however by taking care not to fold in rainy time, you may, on such ground, carry on your folding, till the barley is five or six inches high, and near spindling, with great success.

But,

Chap. III. FOR BARLEY.

But, for further enriching barley land, some, that fat their sheep on turnips, will every day dig up those which have been scooped by them, and feed their store-sheep with them every night in the fold, which these sort will greedily eat after having been kept all day on a short bite of grass; and thus such ground will be three times dressed, once by the fat sheep, next by the store sheep, and lastly, by the same after the barley is sowed; and indeed this is the best of dressing, not only for making the corn stand fast against storms of wind and rain, but also for causing a great crop, and preventing its being damaged by worms.

But, before I quit writing on this chalky soil, I have to observe, that turnip seed is often sown on this earth to profit; and when a full crop of them happens to be eaten off with sheep, cows, or oxen, then they give it only one ploughing, and sow three bushels and a half, or four of barley, on the rough ground, and harrow it in; then directly clap the fold on the same; and, if they can run it over the whole crop, they hardly ever miss a rich return at harvest, because the feet, dung and stale of sheep agree better with short chalk, than with any other earth whatsoever, by their treading in the seed, and plaistering it, as it were, down so fast, that the crops of barley will stand upright, when others that go without this benefit will fall, be laid flat, and, perhaps, half the grain spoiled, as is often the case, especially where the chalk is of a dry and very short nature.

CHAP. IV.
SOWING BARLEY ON NATURAL GRASS.

A Gentleman who held a large farm of his own in the vale, having a meadow that did not produce that profit, which he thought it would do, if the ground was employed under crops of corn, came to a resolution of ploughing it up. Accordingly in *March* he ploughed it up with the foot-plough, and a fin of iron fixed in its share, thus: he began to plough it into a broad-land of nine steps wide, by first making a low ridge and ploughing on each side of it till he extended his ploughing to the before-mentioned dimensions; and as he ploughed (as we call it) round the ridge, he ploughed the grass turf down as thin as possible; and as soon as he had turned down one furrow with turf, he directly with the same plough turned a furrow of mould on the same as thick as he could, so that the grass had no chance to grow through it. When he had ploughed all the meadow up in this manner, he
harrowed

harrowed in barley-feed, which produced the biggeſt of crops. Thus he not only enjoyed a barley crop without loſs of time, but prepared, at the ſame time, the ſame ground for a wheat crop to be ſown at *Michaelmas* following; for the mould was laid on the turf ſo thick, that it rotted the turf by *October* next, when the barley ſtubble was ploughed up, and wheat harrowed in. I ſhould have ſaid there were two foot-ploughs employed in this work at a time, one followed the other; the firſt turned down the turf, and the other turned the mould on it.

CHAP. V.
OF THE SORTS OF BARLEY.

OF all white grain rathripe barley is the firſt that is fit to cut; and what I call rathripe barley, is *Fulham* barley, or *Putney* barley. This is the ſort of barley that is firſt ripe, and fit for mowing; by ſowing this a little later than the common barley is uſually ſown (which it will admit of beyond all other barley-ſeed,) the farmer avoided falling this year into two great misfortunes, that attended all or moſt of thoſe, who ſowed their common barley-ſeed the latter end of *February*, or beginning of *March*, which occaſioned the loſs of many crops of grain, as I have

have before remarked. Then those farmers that sowed this rathripe barley after them, had the pleasure to see their crops grow to their great profit: for so it happened, that some had this sort of barley ripe sooner than their wheat, and got into the barns in the greatest perfection, before they inned a sheaf of wheat, and before thousands of farmers got their common barley crops mowed; which, as I said, occasioned the spoiling of vast quantities of this necessary grain; and that which was saved tolerably well, made but a bad malt.

The *Fulham* barley in the southern parts of *England*, in a hot summer and warm soil, is commonly fit for the scythe two or three weeks before any other, and in ten from sowing time. If a *Chiltern* soil is well prepared for it by several ploughings and a good dressing, it may be sown in *March*, to be ripe in *July*; but, if a turnip-ground, it may be sown in *April*. I have known it sown in *May*, and yet sooner ripe than our common *Hertfordshire* sort. Rathripe barley has a great advantage of all others, for, by its quick growth, it is the less time abroad, and therefore less exposed to great rains, which are enemies to the colour of this grain; and this is the reason why the *Chelsea* and *Fulham* farmers are thought to have the whitest,

most

most thin-skinned, and mellowest barley in *England*, from off their sandy loamy land, that always fetches the greatest price for seed and malt. I never knew barley in general so white as it was in the dry hot summer of 1741, because little rain fell, and it was early ripe. To have barley of a pale colour and thin skin, a particular farmer, in our parts, observes to begin mowing his barley, when he perceives the small black veins or streaks are gone off the kernels, and the ears brownish and half bent. Others more commonly observe to let it stand till the ears hang quite down, close to the straw, before they mow it, and take this for an indication of its being full ripe. About *Ware*, in *Hertfordshire*, which lies near twenty-two miles from *Gaddesden*, they make abundance of malt for supplying the *London* brewer, and are as careful of getting in their barley as they are of their wheat; and it is here that they commonly cut down their barley before it is full ripe; because they are of opinion the kernel is thinner skinned than if it stood longer, the colour paler, and the quantity of flour the same; that the chalky, gravelly, and stony grounds produce the finest skinned and coloured barley, but that the deep moulds on a clay bottom, and stiff loams, yield a thick-skinned sort of a reddish colour. With us we reckon the riper the corn the more flour it contains, and that

moderate

moderate showers, a little before mowing, plump the barley, and cause it to thrash the easier out of the ear. Generally we begin mowing our barley after the wheat is got in; but, in the year 1740, barley was ripe as soon as wheat. I knew a farmer that formerly sowed *Fulham* barley 24 miles distant from *London*, and commonly had it so early ripe, thin-skinned, and pale-coloured, that men from about *Windsor* used to come to *Hempstead* market, and buy his and other such, for making it into what they sell at the grocers and druggists shops in *London* for *French* barley.

Fulham barley will obtain a larger body, and better colour, than any of an intire loam or clay: and as this sort of barley is the soonest ripe, and fit to mow even sometimes in ten weeks time after sowing, I have sent this, and sprat-barley, far into the north, as well as into *Norfolk* and *Sussex*. For, by sowing such a feed, you will not only stand a chance of getting a crop of it into the barn sooner than your neighbour, but the malt, made of such barley, will, by far, exceed, in goodness, any other sown in your country: this I aver for truth, as I have, as well as my neighbours, experienced it: and indeed we not only sell the barley, when made into malt, but we sell the seed of this *Fulham* barley, of the first and second years growth, generally for two shillings

lings a quarter more than our common barley. And many farmers are glad to have it at that price, who have not the conveniency of sending for it. Here then are several advantages attending the sowing of this rathripe barley seed; as its growing sooner into a crop than any other barley whatever, with a thinner and whiter skin, fetching a greater price than ordinary for seed, or in malt, is a beneficial change to the ground; and therefore will pay the sower of such seed to a great degree of profit, if he sow a good quantity of land with this delicate sort of grain.

By some, barley is sown the beginning of *May*, oftener out of necessity than choice: As when the farmer, by multiplicity of business, cannot get his land into a fine tilth time enough, or that he has not eat his turnips off before now, or is retarded through the inclemency of weather. In all which cases, the rathripe, or *Fulham* barley, is now the properest to sow either in vale or *Chiltern* lands, because it grows quicker than any other sort; for, though it is sown late, it is early ripe, and will prove the best of barley for malting, after being once sown in loams or stiff lands, which gives it a much larger body than the sandy ground it came from.

This rathripe barley will acquire, in a stiffer soil than where it came from, a larger body than the
seed

seed had: it will even in such a stiff soil carry a thinner skin than common barley will. The rathripe barley will be ripe sooner than any other barley whatsoever; and therefore has a much better chance to be got into the barn in a drier and finer condition than any other sort of barley. It will sell for more at market than any other. And lastly, it will make better malt than any other sort; provided it be only of one year's growth in such stiff soil, from the time of sowing the true rathripe barley-seed: not but that the virtue of this rathripe barley will remain two years, or more; but then the second sowing of it will not produce such good effects as the first.

Sprat barley is sown in wet and dry grounds; about *Eriff* in *Kent*, they sow it in their rank marshes, because it is more hardy, and will not run into straw like the common sort; but I have seen it grow in drier ground about *Bridge* in that county, where they sow it in such land the latter end of *February*: three bushels of this sort will sow an acre, when four must be used of ours, because it has a two-rowed flat larger ear that requires more room. This barley will stand upright, as having a strong stalk, when ours will fall down, and therefore is more conveniently mowed. Its skin is somewhat thicker than ours,
but

Chap. VI. OF BARLEY. 319

but the kernel rather larger, is harder to thrash out, and longer making into malt, but better for the distiller's use, as yielding more flour and spirits than others. It makes good pale malt drink, and better bread than ours, when it comes off a dry soil and is inned well; for with us we generally sow the common barley, degenerated from the *Fulham* sort. There is another kind I have seen, that is a large-bodied barley, and grows as such in a bearded ear; but its naked kernel looks like wheat, and is certainly the best of barley for making malt, &c. The four-rowed big barley we sow hardly any of, in the southern parts of *England*, because it is a smaller grain, and makes coarser malt and bread than ours. But in the northern parts, it is much sown for its hardy nature, and growing on poor land.

CHAP. VI.

OF THE TIME OF SOWING.

ABOUT the 12th day of *March* is the usual time that vale-farmers begin to sow barley, if the weather is in their favour; and this time they more strictly observe of late in *Aylesbury* vale, than they did formerly; because many of

them

them have suffered very much by sowing their barley seed sooner, in hopes to have their crop ripe before their neighbours, and get it into their barns in a dry season, and with a fine pale colour. But the consequence of too early sowing barley-seed has often been attended with very ill effects; for, if rains fall for some time quickly after sowing, and cold weather happens at the same time, the barley-blades will discover to the owner the poor condition its roots are in by their reddish fox-colour, which is a certain sign of a lamentable poor crop at harvest. And, if a vale farmer should defer the sowing of his barley-seed too late, then he is in danger of suffering by a bad crop another way; that is, if a long dry season succeeds such late sowing, then the crop is the more likely to be attended with the foregoing misfortunes of growing in three degrees of ripeness.

I knew a person plough up one part of his turnip-field, as soon as the turnips were eaten off, and sow the same on one ploughing with barley in *January*. And in *February* he did the same by another part of the same field, as soon as the turnips were eaten off; and in *March* the like by a third and last part of the same field. Now this he did to try a practice, different from any in the country; and although the soil of the field was a wettish loam, he had the best crop of the three,

from

Chap. VI. OF SOWING. 321

from that feed fown in *January*; and that fown in *February*, was the next beft; and that fown in *March*, the worft: and all this happened by means of a very dry fummer that followed; by which, the two laft fowings of barley were deprived of that moifture, neceffary for the production of a full crop of this grain. But this was owing to the accidents of weather; for, in fome years, the late fowing of barley proves as much for the better, as fuch early fowing did this year. I have known a rafhy, fharp gravel, fown with barley after turnips, in *April*, and yet the owner had an extraordinary great crop, by the help of the following favourable wet weather.

The vale farmer thinks he runs a rifque of the leaft danger, when he fows his barley about the twelfth of *March*; for though, by this early fowing, he is under the more danger of being hurt by floods and froft, yet is he alfo entitled to the hopes of a moderate, moift time afterwards, that may bring on fuch a timely head of barley as to cover and fhade its roots foon enough to withftand a dry fummer.

Vol. I.   Y

CHAP. VII.
OF THE QUANTITY OF SEED.

ON this account, I shall observe, that the quantity of seed ought to be proportioned to the nature of the soil, tilth, and the season of the year. *First*, as to the soil: On stiff loams and clays, some will sow, in the common way, four bushels on one acre; because on this surly, cold glebe, the seed is very apt to bury, chill, and die; and therefore such land should be assisted with good warm manure, as will enable the barley to grow thick, and kill the weeds, which, in such sour ground, are very apt to get rampant, and destroy the corn. On the contrary, in many of the light, sandy grounds of *Norfolk*, and *Suffolk*, they sow but one or two bushels on an acre, and if they increase the quantity of seed, it is as the land is heavier: because, as they say, their light, poor ground is not able to carry more; for, if they were to sow it thicker, such land would return an hopper-eared crop at harvest, or in plainer *English*, a little ear with a few kernels.

Secondly, If a stubborn soil happens to lie in a sour tilth, as it is very apt to do, there must be the more seed sown, to allow for a considerable loss, by the cover that the clotty part of the earth will cause, and so deceive the farmer, if he he allows not enough to prevent this misfortune. On the contrary, when land is in a fine tilth, the less seed will do; and this is one reason why in sandy grounds they sow so little seed; for that in such pulverised tilths, hardly one kernel misses coming up. *Thirdly*, If you sow late, there must be the more seed sown; for it is a standing rule in husbandry, that, the later you sow any corn, the more seed should be allowed; because then the grain will never gather, branch nor kern so well as the more forward sown, as being drawn up so fast by the powerful attraction of the sun, that it will run a-pace into stalk, and less into corn; and because the fowls, in late seasons, live mostly on the grain sowed then, as having little elsewhere for their subsistence; and here they have sometimes a long opportunity to scratch with their feet and dig up with their beaks, the barley corns, before their blade is seen above ground, when a dry summer directly succeeds the new sown grain, to the destruction perhaps of half the late sown crop.

CHAP. VIII.

OF THE CHANGE OF SEED.

THE change of seed is of such importance towards improving crops of grain produced by it, that many send for wheat, barley, pease, &c. to great distances. For this purpose, there are many sacks of wheat carried out of *Hertford* and *Bedfordshire* into *Northampton, Leicester,* and *Darbyshire,* for the sake of sowing a seed that comes off a chalky soil, in a clayey or stiff loamy soil; so the *Hertfordshire* farmers, several of them, send for *Fulham* barley-seed above thirty miles an end, and all by land carriage. Now, though we have sandy, chalky, and gravelly lands just by home, yet we of *Little-Gaddesden* chuse to be at the extraordinary charge of sending for this *Fulham* barley-seed, though we live thirty-four miles from it, and find our account in so doing; for as we sow it in our stiff loams, from off a sandy short loam, it returns us a very early crop, with a kernel much bigger than that we sowed, and is so natural for making true malt, that it is commonly sold for two shillings a quarter more than our common barley is; but there are other reasons for our preferring this *Fulham* barley-seed before all others. One is, that by getting a crop of such barley, sooner than ordinary, off the ground, it gives a farmer the opportunity to sow the same land with turnips, early enough to enjoy a full crop of them. Another reason is, that we can (if the land is proper for it) sow rape-seed. Another is, that this barley being ripe before wheat,

wheat, it may be very probably got into the barn in the driest order, as theirs about *Fulham* commonly is, who thereby are masters of the best of barley-seed. And it is this excellent quality that makes it a fit sort to be sown in the northern parts of this kingdom, because this barley-seed, beyond all others, may be sown late and mowed early. In short, there are these two great conveniences attending this peculiar sort of barley. The other is, that, notwithstanding such late sowing, the crop, produced from the seed of this *Fulham* barley, will be ripe as soon, if not sooner than the wheat-crop.

CHAP. IX.

THE GROWTH OF BARLEY.

BARLEY-crops likewise, in the year 1744, suffered very much, by three several accidents of weather, especially that which grew in *Chiltern* countries: first, by snows; secondly, by droughts; and lastly, by rains. First, by snows, that began to fall on the 31st day of *March*, 1744, and continued snowing the greatest part of the time between that day and the fifth of *April* following; so that the snow lay several inches deep in the snowing-season, which disappointed some from making an end of sowing their barley-

barley-feed, and others from beginning to sow it till very late; for after the snow had fell, rainy weather succeeded, and obliged many to defer ploughing and sowing till the 23d of *April*. However, some adventurous bold farmers, that had a great deal of ground to sow with barley, run a risque of the weather, and went on ploughing and sowing of barley in their dryish soils, in such snowy weather, but paid dearly for it; for they who proceeded in this manner, had not above half a crop of barley at harvest; for snowy and wet weather is a great enemy to good ploughing and sowing. Secondly, barley-crops very much suffered this year, by reason of a long dry season, that presently succeeded the sowing of the seed; which caused that seed which lay lowest in the ground to sprout, and come up first, from the larger share of moisture the greater cover of earth occasioned. The next that sprouted was that barley-seed which lay somewhat higher: and the last, that which lay the nearest to the surface, as having the least share of moisture, and cover of earth; whereby were produced the several gradations of the seed's growth, and consequently several degrees of ripeness at harvest; that is to say, ripe barley, half-ripe, and unripe, or green barley; as was the very case of many thousand acres of barley this harvest: yet these must be mown, and mixed together; and in this condition,

dition, malt muſt be made of the ſame, after the barley has paſſed through its ſeveral degrees of ſweating in the mow. But what ſort of beer and ale muſt ſuch malt make?

CHAP. X.
OF THRASHING AND CLEANING BARLEY.

TWO men are more neceſſary to thraſh out barley than any other large grain; for there is no corn, that I know of, ſo liable to withſtand the ſtrokes of the flail, as barley, and this for the following reaſons: Firſt, when barley-ſeed is ſown late, which the farmers in many places are forced to do, becauſe of eating their turnips or cole off ſo late as *April*, it often happens, that dry weather ſucceeds the ſowing of it, and ſometimes continues four, ſix or eight weeks together. In this caſe there generally enſue two or three ripeneſſes in a barley-crop. Now ſuch a medley of ripe and unripe barley gives the thraſher occaſion to exert his ſtrength, and try his ſkill to the utmoſt; for, as there are not only many unripe corns in the ears, but alſo many ſmall dwindling half kernels, that ſtick ſo faſt, as hardly can be got out with all the labour that can be uſed by the thraſher, there muſt be conſequently fequently many left behind; but the more where only one man is employed in thrashing, as I shall by and by further observe. In the second place, when ground is dressed or manured in such a rich manner, that, with the help of such a rainy warm spring and summer, the barley runs into so great a crop, as to be laid before harvest, it then commonly yields a lean thin kernel, that will be difficult to get out. Thirdly, it always happens, that, when barley is mowed before it is full ripe, it proves very hard to the thrasher to force it out of its ear. Fourthly and lastly, if barley is thrashed before it has done sweating in the mow, it adheres very close to its hose, and must require hard labour indeed to thrash it clean out; because it misses one part of that maturity or cure, that is requisite to make it part easily from its hose and ear.

As to the first and second, there wants no advice; and, for the third and fourth cases, he must be an ignorant, or a knavish farmer, that suffers his barley to come under these misfortunes. Why I make use of the term knavish is on account of those, who, to preserve a fine pale colour on the barley-corn, thrash and sell it presently after it is out of the field, without giving it time in the mow to sweat, left the weeds that may be mowed and carried into the barn, with the barley, cause it to sweat so much, as to

give

Chap. X. CLEANING BARLEY.

give it a higher colour than the farmer defires. When it fo happens, the brewer commonly fuffers, becaufe fuch unfweated barley-corn feldom fprouts and throws out its fpires in a right manner for making true malt: however, let barley be got in ever fo good order, two thrafhers at a time, in one barn, are far more preferable to the owner than one; becaufe two flails follow one another with fo quick a motion, that they keep it from jumping about, and confine the ears in the way of its ftrokes much clofer than a fingle flail can; but the main reafon why two flails are better than one in this work is, that they can beat and break the ails or beards of the barley much more than a fingle flail can: a thing fo neceffary to be done, that, if barley is not delivered from thefe exuberances, it will fetch the lefs money at market; for thefe ails ferve only to fill the bufhel, and do no fervice to the buyer. After barley is thrafhed out of the ftraw, it is for the moft part thrafhed over again, in order to tail it, or, as I faid, to get thefe ails as clean off as they can; and, after all their labour in thrafhing of the ears and ftraw with the flail, there remains fomething to be done befides to clear the barley of its hils; and that is, as foon as the ftraw and its ears are well thrafhed, and carried off the floor, then, if the barley crop was inned in good order, the two thrafhers will

work

work on with their flails, and foundly beat the barley again, while it lies in its chaff, to break off thefe tails or ails that may remain on it; but, if the barley-crop was inned wettifh, or otherwife ill, they commonly throw it firft out of the chaff, and then flack the heap of corn not only once as it lies, but they turn it and thrafh it again and again, and all for clearing the barley of its pernicious ails or tails.

Of Cleaning of Barley.

In this and in moft other countries of the *Chiltern,* they differ from the vales, in the manner of cleaning their barley. When it is thrafhed out, and the ails are beat and fufficiently broke, they pafs the corn through a large-holed caving fieve, for taking away the fhort ftraws and offal trumpery, in order to prepare it for cleaning, by throwing; and where there is room enough, two men may throw it well to clear it of its chaff, the feeds of weeds, and the lighteft corns. If the barley was inned pretty clear of weeds, was full ripe at mowing, and had its due cure, by fweating in the mow, then, perhaps, once throwing will be enough to make it ready for a further cleaning, by the great wooden and wired fcreen; but, if it was ill inned, then it commonly requires a fecond throwing, to take out thofe light and lean kernels that would not

drop

Chap. X. CLEANING BARLEY.

drop short in the interspace at the first throwing. When the barley is throwed enough, the next thing to be done is, to ridder or riddle it; for which purpose, a boy attends the man who sieves it, by supplying him with more barley, as fast as it is discharged by the sieve, and so on till all is done riddling, and made thus ready for the great screen; I say, a wooden and wired flat screen that stands a little sloping, the better to let out the dust, seeds of weeds, and light kernels, and to distinguish it from the long wormlike wired sieve; by which both wheat, barley, and oats may be very expeditiously cleaned in great perfection.

When all the barley is thus screened, it is ready to put into sacks for market; but, in case some seeds of darnel should be left among the barley, as it often happens, because these seeds are so near the bigness of a barley corn, that they cannot be easily separated, it is not of great importance or damage to the barley; for that this seed is of such a nature, as to add a strength to the liquor, and make the beer or ale, brewed from such malt, the more potent. But the main matter is to free it of its ails or tails; for, if these are left, in any quantity, among the barley, there will a lower price be bid for it, than others that are free from such trumpery. It is this that makes the cleaners of barley oftentimes walk on

the

the thrashed barley, to tread on and break off these ails; and sometimes neither thrashing nor treading will thoroughly break off these ails enough to make the parcel intirely clear of them. One man will sometimes thrash and clean six or eight bushels of barley in one day; and sometimes it is work enough for two. But when barley has been inned well, and a plentiful crop has happened, one man has cleaned five quarters of it in one day, I may go farther and say six, but then it was all throwed out of its chaff before he began cleaning.

The Vale Way of cleaning Barley.

As to the thrashing part of barley in vales, it is done the same as in *Middlesex* and *Hertfordshire*, as I have before observed; they only vary in cleaning it; and that is thus: after they have thrashed out the barley clean from the straw, they rake the coarse offal short straws off, and then pass barley, chaff, and all through a sieve, to take out the next offal short straws and ears that may remain among the corn and chaff: when this is done, they bring the wind-fan near the heap, and then with the help of a man to sieve, and a boy to supply it, they fan as the barley runs out of the sieve, and blow away the chaff from some of the lightest kernels into the yard; and for this purpose they lay the heap of barley, and place

the

Chap. X. CLEANING BARLEY.

the wind-fan fo near the barn-door, that it may be blown into the yard for fattening their fwine, and feeding their poultry. This method of wind-fanning corn is not only practifed in vale countries: it is alfo done much in the weft country among their dry farms, and they take this method to fatten their fwine in part for bacon; for a porker will fat alone upon the light kernels, and other trumpery that is blown out of the barn; for here they take all the care poffible to fatten a great number of fwine every year; becaufe almoft their whole living depends upon this fort of flefh, as is evident in fome meafure by the common anfwer a traveller meets with at their inns, when he afks what is for fupper? Their reply is, Pork and cale, or cole, or more properly coleworts. Accordingly their butchers make no difficulty of killing bacon-hogs and porkers throughout the whole fummer in many places of the weft. At *Frome* I faw it done, I think in *July*, without fearing any damage accruing by the great heat of the weather, which in many other places proves a fufficient reafon to hinder the flaughtering of hogs, left they fhould not make the flefh take falt, and keep fo fafe and found as it will in cooler weather: but here, I fay, they are not apprehenfive of any fuch danger; becaufe, prefently after killing, they throw the flefh into a tub of brine. But to return to
my

my subject. After the heap of barley is once cleaned by the wind-fan, and they perceive it will be to their advantage to fan over again the lightest part of the barley, they do it a second time; and then the wind-fan leaves all the corn in so compleat a condition, as makes it fit to be put up in sacks for a market-sale. And thus two men and a boy will clean more quarters of barley in a day, than any two men can by throwing, riddling, and screening in our *Chiltern* country; and, at the same time, clear the barley from soil and trumpery far better than ours can.

But, although I praise vale-farmers, and others, for their more expeditious and cleaner way of preparing wheat, barley, and other corn for market, than what we do in the *Chiltern* country of *Hertfordshire* and in *Middlesex*, yet I cannot well help dispraising them in their extravagant oeconomy of fanning their light corn and chaff into the farm-yard, because they must in course waste and lose a great part of their chaff.

CHAP. XI.
OF HARVESTING BARLEY.

OF this management none are more curious than the *Chelsea* and *Fulham* men, who, in case their barley is mown in a very dry hot time, will sometimes cock it up in a dewy morning, on purpose that it may have a little and quicker sweat in the field, to preserve its white colour the better, and soften its kernel; for too great a sweat in the mow is apt to make it reddish or black-coloured, and sometimes give a musty smell; and thus their barley looks white, bites mellow, has a good body, and smells sweet in the highest perfection that ever I saw any. On the contrary, if the summer proves wet, and clover or other grass or weeds grow thick among it (as they too often do) the barley must lie abroad, till, as I said, they be dead dried, else it will acquire a red or dark colour by mow burning, which is of such a great prejudice to barley, that it often hinders such seed from growing, and then, I am sure, it will be unfit for malt. Likewise if barley is grown in the field, it loses so much of its farinous vital property as can never be removed. When either of these ill qualities happen to barley in any great degree, it will be best employed in feeding horses, hogs, or poultry.

It is a general way throughout *Hertfordshire*, and all other counties that I have been in, to rake and cock up barley, which we in a loose manner carry home in a cart or waggon to stack or mow; but this loose method is refused in *Kent*, where they are celebrated husbandmen, and where, after their barley is mowed down, with the scythe and cradle, they make use of a sort of rake, with five iron or wooden teeth in it, three of which are six inches long, and two nine inches long; with this a man rakes up a parcel of barley on each side of him into a heap, to be bound up in a bundle, with some of its own straw, by another man that follows, and so goes on throughout a field. These bundles they let lie in several heaps to be loaded in carts or waggons. Now, for justifying this procedure, they told me, they would wager, that they would carry in a field of barley sooner in this way than in the loose common way; and when they had it in the barn, they would also wager, that a man could thrash more in a day of it, than another could after the usual loose method, because by this means they could lay all the barley ears against one another, as we do wheat sheaves.

CHAP. XII.
PRODUCT OF BARLEY.

WHOEVER ploughs his ground often enough, whether it be a gravelly, chalky, or loamy foil, and dresses or manures it well, and sows it in a dry season, need not fear, by the blessing of heaven and these means, six, seven, or more quarters from an acre. I have known nine quarters to grow on one acre throughout a field of nine acres. A marsh-soil, in *Kent*, has yielded eleven quarters off one acre of sprat-barley; and yet in some of their clay-lands in *Middlesex*, they seldom get more than two or three quarters off an acre, because they sow the wrong barley in a wrong manner. In the year 1737, I had several roots of barley that seemed each of them to produce thirty-seven stalks, and their ears had twenty-eight kernels each, one with another, in all 1036 kernels from one root.

BOOK VI.

Of the CULTURE of OATS.

CHAPTER I.

OF THE SOIL FOR OATS.

THE soil for sowing white oats is of the greatest importance to be considered; for, if this seed is not sown in a proper earth, the labour, money and time, will be in a great measure lost. I knew a *Chiltern* farmer sow white oats in a wettish loam, that returned him indeed straw enough, but the oats were of so thin a body, that they were next to chaff; which so discouraged this farmer, that he never afterwards sowed any more white oats, though his large farm contained various soils. On the contrary, I knew a farmer that lived at *Bragenham*, near *Leighton*, in *Bedfordshire*, every year sow no other oat, than the white sort, because they returned him

him prodigious crops off his sandy grounds, even to nine quarters off an acre, and that of the largest sort of kernels; but then he kept his land always up in good heart, by the help of the dung of many cattle, for this farmer employed both grazing-ground and ploughed ground, which gave him an opportunity to keep abundance of cattle, and them of returning him abundance of dung. Now this farmer was even necessitated to sow white oats, because this dry ground will not bear good wheat, nor good barley; therefore he chiefly sowed rye, turnips, and white oats; and, for obtaining a large crop of such oats, he never failed of making choice of the largest and cleanest seed; and it is by these means he used to sell the best of oats from off a poorish earth, as his naturally was; but, when such a soil of poorish land is enriched by the dung of animals, or by manures, it then in many parts of *England* agrees excellently well with barley and white oats; as is obvious to all who travel through the counties of *Suffolk, Norfolk, Kent,* and other parts, where, if they have judgment enough to make due observations in their way, they may see great quantities of sandy land sown with these two sorts of grain, as those which best agree with it; not but that a white oat will grow very well on gravels, chalks, and dry loams.

But the chief miſtake of all others is, the ſowing them in ſuch ſoils that are ſtrangers to the dung-cart, or other fertile dreſſing : for to expect a full crop of white oats off a poor barren ground, is the ſame as expecting a horſe will do as hard a day's-work, with part of a belly-full of meat, as thoſe who enjoy a whole belly-full; yet, as abſurd a thing as this is, it is the caſe of many ignorant farmers, who truſt ſo much to a wet warm ſummer for producing them a plentiful crop of oats, that they even are careleſs about the matter; and the more, becauſe oats will grow into tolerable crops, when hardly any other corn will, with little or no dreſſing; but he who dreſſes his ground on purpoſe, or he who otherways keeps it in good heart, has two to one the better chance of ſucceeding, as I am going farther to obſerve.

Dreſſing, as I ſaid, or not dreſſing ground on purpoſe for obtaining a full crop of oats, is not worth diſputing; the matter of fact is, whether the ground is in good heart, at the time of ſowing the ſeed; if it is, an oat-crop may, perhaps, pay as well as a wheat or barley-crop, in ſome particular years, when the two latter are cheap and the oats are dear, as it ſometimes happens to be the caſe; for ſix or ſeven quarters of oats, at ſixteen ſhillings a quarter, may fetch as much money as twenty-five or thirty buſhels of wheat will, at three ſhillings and ſix-pence a buſhel.

CHAP. II.
OF PLOUGHING FOR OATS.

I Know some *Chiltern* farmers, who dare not sow some of their fields with oats, unless it has had two ploughings, to give the oats room to grow up, and cripple the weeds. These ploughings are also most necessary to be performed, in case a farmer sows clover, or rye-grass, or trefoil, or sainfoine-seed among his oats; for, though many are so silly to sow these seeds on only one ploughing of the ground, it often causes them to lose their grass-crop by it; whereas had the same earth been ploughed twice, and got thereby into a fine tilth, such grass-seeds have three times the chance of taking the ground; and, indeed, I do aver it for truth, that an oat-crop is the properest corn of all others, to sow any of the grass-seeds amongst, if the ground is in good heart; because the stalks of oats are apt to stand stiffer than barley, and thereby the crop of grass is in less danger of being spoiled. When the ground is ploughed ready for sowing of oats, they are to be sown half on the rough earth, as the plough left it, and then the harrows are to follow; and when the ground is harrowed once in a place, the other half of the seed is to be sown broad-cast like the last, and harrowed in long-

long-ways and crofs-ways of the land, till the feed is well covered; but there has been a practice carried on by fome particular farmers that I have known, who think it the beft way to fow oats by a fingle throw of them over the whole ground, as believing they can, at fuch a fingle throw, fow feed enough; and if a perfon tells them, they cannot fow the ground thick enough in this manner, the anfwer I remember of one was, that his hand was a large one, and therefore could compleatly fow the ground at one throw or broadcaft; but their affurance has fometimes failed their expectation, and found the contrary effect, by the oats coming up fo thin, as to give the weeds room to grow between them, to the great damage of the crop. For my part, I always have my oats that are fown in the broadcaft mode fown twice in a place, and find it far the furer way to have a full crop, than fowing them by a fingle thorough; and I have tried both ways.

CHAP. III.

OF THE TIME OF SOWING OATS.

THE beginning of *February*, and fo on, plow up your land for oats; and, from the middle to the latter end of *February*, fow and harrow

Chap. III. OF SOWING OATS.

harrow in black-oats. They are generally sown on one ploughing, after wheat or barley: formerly, seldom till *Lady-Day*, or the first week in *April*; but the present practice is to sow them all *February*, as well as all *March*, but rarely after. Oats will thrive the better, if they have a good fine tilth to be harrowed into it, but they are seldom allowed it; otherwise than that the frosts shatter the surface of that ground, which was ploughed up for oats in the preceding *November*; and, by lying till *January*, or in this month, its top-part becomes loose, and sweet for oats to be harrowed into the same. It is a very hardy grain; and those, that sow them forward, will not only stand the best chance of rain to bring them forward, but they will corn the better; for, if oats are later sown, it is a chance if the dry weather does not lessen their crop; and, if wet weather follows quickly after such late sowing, it is a chance if they do not run into much straw, and less corn.

February is the best month in the year for sowing the black oat in the general; but I have known a farmer, whose soil was of the chalky sort, give it only one ploughing with a foot-plough, and plough his oats in in *January*. Thus, by sowing them under thorough, and a month earlier than is usually done, he said, he would not be catched as he was last year by a dry

dry summer, by sowing his oat-seed in *February*, and he succeeded accordingly; for he had a great crop, though a dry time followed his sowing the black-oat in 1740. But the case is altered, where a moister ground is to be sown with oats; for, if such a soil was to be sown in *January*, and a hard frost happens, the new spiring blade will be in danger of being killed; yet I knew a great farmer venture at sowing a gravelly clay with blacks-oats likewise in *January*, and he had seven quarters of oats in return on each acre; but it was chiefly owing to a very mild spring, and dry summer that followed; however, this was a chance hap, and is not to be presumed upon in the general way of sowing oats; therefore *February*, or the very beginning of *March*, is the only safe season for sowing the black-oat.

The various natures of oats contribute very much to a farmer's conveniency. The black oat is reckoned the hardiest oat of all others, and therefore is sown sooner than any, and so affects an early sowing in *February*, that that month is accounted the best month in the whole year for sowing it. On the contrary, a white oat is of a more tender nature, and more forward or quicker growth; which obliges farmers to sow them later even in the months of *April* or *May*. Now as a white oat requires a later sowing than a black oat,

oat, it gives many *Chiltern* farmers an advantageous opportunity to feed their turnips, their rye, or their rapes off late, and yet not be deprived of enjoying a good crop of corn at the following harvest: for when any of those are fed off by sheep, cows, or oxen, in this or next month, white oats may be sown to a good purpose; and it is by this means of feeding turnips, rye, and rapes off late, that either rath ripe barley, white oats, or blue, or other tender pease may be sown in this month, directly after such a crop is got off; when it is an unseasonable time to sow black oats, because they will run into straw or stalk much, but little into corn. But white oats, having a very strong, large, flaggy stalk, and ear of a great bigness, employ the fertility of the earth for nourishing these its branches to that degree, as to hinder its too luxuriant growth, and prevent the violent attractions of the sun's heat, from lessening its multiplying into plentiful crops of corn, when the more naked stalk of black oats, if sown so late, would be overcome by them, and be rendered a poor crop; because the roots of the white oat grow under a sort of cover or shade that they receive from the green broad blades that always accompany their stalks; which leads me to make observations

on

on that earth which is most proper to sow white oats in.

The black-oat is for the most part sown in *Hertfordshire*, and other *Chiltern* countries, for the particular reason of its ready growing and thriving in ordinary lean soils; for in *Chiltern* countries we do not pretend to vie with vales for goodness of earth; and therefore when they sow oats (which is but seldom) they sow the white-oat for the most part, that requires the best ground to plough in, as I shall more particularly observe in the next article. A black-oat has a lesser body than the *Poland*, or white-oat, and has but two hulls or skins, where the white-oat has three. A black-oat, by some is accounted a sweeter oat than the white-oat. Others are of a contrary opinion. In mine, I think no oat exceeds a good black-oat for making the sweetest of oat-meal, and as white, or whiter, than that of a white-oat. I have sown a white-oat, but left them off for the following reasons: Oats are commonly sown in *Hertfordshire*, and most other countries, without any dung or manuring the ground for them, as being a *Lent* grain, that most commonly are sown after a crop of wheat or barley.

Of Sowing Oats in March.

Though *February* is the properest month, of all others, to sow oats in; yet there are thousands of

Chap. III. OF SOWING OATS.

of acres fown in *March*, efpecially where the foil is of a clayey nature, or lies in a wet fituation; for in cafe even a gravelly, chalky, fandy, or loamy foil lies very near the fprings, or is otherwife fubject to fuffer by water; fuch ground, I fay, is better fown with black or white oats in *March*, or *April*, than in *February*; for, as hardy a grain as this is, it may be chilled and killed, where the feed is fown too foon in land expofed to inundations of waters; for which reafon hundreds of farmers, in *Aylefbury* vale, dare not fow any oats in that ground, that is of a clayey nature, and lies in fuch low fituations as fubjects the corn to the damage of waters: therefore unlefs the foil is of a dry nature, no fort of oats ought to be fown in *February*; and, if it is of the wetteft fort, then *March* affords the farmer the fafer opportunity.

And now may be fown either the black, or white, or reddifh oat, with or without any dreffing or manure; for an oat will profper where hardly any other corn will, for which reafon, the oat is become the chief, and the moft profitable corn for fowing and profpering in the remoteft parts of the north, where fome fow the reddifh, others the white, and others the black oat. But the reddifh oat, which is much fown in *Derbyfhire*, gets more and more into reputation for an extraordinary good oat. In *Scotland* thoufands

of

of acres are sown with a smallish white oat, that are not so large as those sown more to the southward. The white oat is the tenderest sort of all others, and therefore ought not to be sown before *March*, or in *April*; and then in a sand, in a chalk, in a dry loam, or in a stiffish earth, so it lies not wet; and when it is sown in any of these soils, and they be in good heart, a white oat will flourish and become a great crop. But I do assure all those who sow this oat, that though all the oat tribe are great peelers or robbers of the goodness of the earth, yet this will do it beyond all the rest; for this oat has not only a strong large stalk and ear to nourish, but also a broad flag besides, that keeps company with the stalks. It is therefore that in most of our *Hertfordshire* grounds, as well as in several other *Chiltern* countries, they sow the black oat, for its great yield and less suction of the earth than the white oat draws; for it is a folly to sow a white oat in a poor soil.

And although most farmers in *Hertfordshire*, and elsewhere, never dress their land directly for sowing it with oats, there are some that wisely dress their ground with dung or compost on purpose to nourish an oat-crop; and this they do very judiciously, because dung, laid on to assist an oat crop, will take off that rankness that would canker, or smut, and spoil a wheat crop

Chap. III. OF SOWING OATS.

in some sort of soils, if laid on the ground just before the wheat is sown; which, by this piece of forward husbandry, is delivered from these mischiefs; for the ground, being thus enriched by dunging, will certainly produce huge crops of oats, and a huge crop of oats will certainly clear the ground of weeds, and leave it in a most fine hollow tilth condition.

It is a very unusual thing, but two farmers would boldly venture to sow their ground in *January* with oats, in order to have their crops off very early; accordingly, they sowed their gravelly clayey grounds with black oats, and had a great return at harvest, for, as it luckily happened for them, the dry summer of 1741 succeeded so well in their favour, that they had seven quarters on an acre of black oats.

A neighbour of mine sowed his oat-seed thin and late in a cold, wettish, flat land, which was a four tilth. The consequence was, he had no more than two quarters off one acre, when his next neighbour, who sowed early on the same soil, had four quarters off an acre. Sow early and have corn, sow late and have straw.

CHAP. IV.
OF THE SORT OF OAT.

THERE is a *Dutch* oat that has a thin skin, a short, plump, white kernel, makes good oatmeal, and more than the common black oat, because of its thin skin, which gives room for the more flour, grows with a reddish stalk when near ripe; and not very tall. It is sown about *Croyden* in *Surrey*. The *Scotch* white oat is also a good hardy oat, but, in *Heriford/hire*, we for the most part sow the black oat, which we reckon has the thinnest skin, and makes the sweetest oatmeal. It is allowed by all that a white oat peels and impoverishes the ground more than a black oat, although either of them do it so much that we are obliged to make a fallow after them, to recover the ground, especially after the large *Poland* white oat.

CHAP. V.
OF THE QUANTITY OF SEED.

THE sowing of white oats deserves a particular regard, for there is this difference between the white oat, and the black oat: the black will gather and branch much more than a white oat, and, therefore, if an equal quantity of of seed was to be sown on an acre of ground, the farmer may be deceived in his expectation at harvest; wherefore, the *Bragenham* farmer I have been mentioning, does not sow less than five bushels of white oats on every acre, in order to allow for the deficiency of their gathering and branching; which is one more than we commonly sow in our loams, when we sow them with black oats; for with us about *Gaddesden*, we seldom ever sow more than four bushels on an acre of these black oats; but then, whether they be white oats or black oats, they ought to be sown twice in a place, over the whole field; for, many obstinate and ignorant farmers have suffered by sowing them only once in place, as trusting to the largeness of a sower's hand, to fling enough out of it at one throw. And such a double sowing is of the greatest importance, for on the thick growth of a crop very much depends the bigness of it at harvest, because by such a thick growth the weeds are overcome and kept down from hurting the oats; and, likewise, the heats and droughts kept the better out from parching up the roots of the oats, which, in too thin a crop, often proves fatal to it; for, when oats are sown in the random or broad-cast way, there is no more mould allowed their roots than what the harrows and roll give them; which, at best, is but a superficial

perficial and moſt thin covering, and, therefore, the more liable to ſuffer by droughts, which is different from the way of ſowing oats in drills.

In ſowing all ſorts of oats, it particularly concerns a farmer's intereſt to adjuſt the quantity of their ſeed, as well as that of any artificial graſs, that is to be ſown among them. On this account he ought well to conſider the nature and ſtrength of his ſoil. If oats are to be ſown in a ſand, or a gravel, or a chalk, without any graſs-ſeeds, there ought to be the greater quantity of oats ſown, the better to enable the green crop to lodge and ſhelter the water of rains and moiſture of dews, for keeping ſuch dry earth in a moiſt condition; but then, if ſuch gravel, chalk or ſand is in ſo poor a heart, that it is not able to nouriſh and bring ſuch a poor crop to a full perfection of growth, it will ſuffer by growing in ſhort ſtalks, and poor lean ears, and thin kernels, and is thus the more unfit to be ſown with clover or any other artificial graſs-ſeeds.

CHAP. VI.
PRODUCT OF OATS.

THIS grain, if encouraged by a ſufficient quantity of dung or other manure, will return great quantities at harveſt. One acre has produced

Chap. VI. PRODUCT OF OATS.

produced ten quarters of white oats. But, of the black fort, our ufual quantity in *Hertfordſhire* is about half fo much without dreſſing, for I never knew any dung for oats in our parts; if they did, there would undoubtedly be feven or eight quarters off an acre of black oats. An oat is faid to root deeper than a pea.

BOOK VII.

Of the CULTURE of PEASE.

CHAPTER I.

PLOUGHING FOR PEASE.

ONE old rich farmer, who I well know to have gotten an eſtate, by his long holding a very cheap farm, and by improving it in many branches of huſbandry, yet was ſhort on one main article; and that was by his harrowing in all his hog-pea-ſeed on the rough ground, which though it would have been abſolutely right management, in a clayey loam, an intire loam, and a gravel, yet was wrongly done here. And this farmer, after holding a farm above forty years, was at laſt convinced, that in a ſhort, looſe, chalky ſoil, as a great deal of his ground was, he ought to have ploughed his peaſe in, and not only harrowed them in on his rough earth, as the plough left it; however, in the year 1742, he ſowed all his hog-pea-ſeed on the chalky land,
before

Chap. I. PLOUGHING FOR PEASE.

before it was ploughed in the broad-caft way, and then ploughed them in with a foot-plough, which covered all his feed-peafe, and thereby defended them not only from the beaks of field-fowls, but alfo a worfe misfortune that oftentimes befalls pea-crops; and that is the drought of long dry fummer feafons, which generally, in this dry, hufky foil, parches the pea-roots, if the feed was only harrowed in. This farmer, therefore, got wife enough to prevent, in a great meafure, this fatal misfortune, though not till the age of feventy-two years; and by the fuccefs that attended this better way of fowing his feed under thorough, he is refolved to continue it, whilft he lives; but, by the way, I am to obferve, that he gave his chalky ground only one ploughing in all, and that was done at the time of fowing his pea-feed; for, if he had winter-ploughed it, for this purpofe, he had been in the wrong of it, as I am going to fhew of this work in other foils.

In the vale of *Aylefbury* moft of their land, that always lies ridge and thorough, is of a bluifh, loamy clay, which, though of a ftiff and very hard nature in the fummer dry time, yet, on a little froft, will fhoal and crumble into a fhort body and furface, and lodge the waters to that degree, if it is winter-ploughed, as to greatly damage either beans, peafe, or thetches, that

may

may be sowed in *January*, or in *February*. Wherefore they are judicious enough to sow these *Lent*-grains on the stubble of wheat, or barley, and then give it one ploughing and harrowing, for compleating their sowing. Thus also in the vale of *Derby*, their farmers are very sensible, that winter-ploughings would do their ridge lands a great deal of harm, because the rains would wash away great part of their goodness, if ploughed twice for pease or beans, and got into so loose and hollow a condition, as to let the air and sun in that would thus dry up the pea-roots.

In the vale of *Derby*, their earth is a reddish clay, of a contrary colour and nature to that in *Aylesbury* vale, and therefore they say, they should lose their crop, if they plough it twice for pease or beans, because, likewise, the waters would have so free an access to, and entrance into, as to make a lodgement in their ridge-lands, load them with it, and cause the harrows to harrow their surface into a marsh after the seed is sown; and the more, if it should rain in the sowing time, to the loss of the crop.

Wherefore, as their ground in *Derby* vale is a red and stiffer clay by far, than that in *Aylesbury* vale, their farmers are obliged to take a different method to sow their pease or beans. In *Aylesbury* vale, as I said, they plough their pease and beans.

Chap. I. PLOUGHING FOR PEASE.

in; but in *Derby* vale, as soon as they have ploughed their land once, they sow their peafe and beans broad-caft all over the rough ground as the plough left it, and harrow them in with one fingle drag-harrow as they there call it: by which mode of fowing, the pea and bean feed falls in between the furrows, and a great deal of mould falls in upon them at harrowing-time, and thus becomes a food and great nourifhment to their crop all the fummer after.

Take care you do not bury your peafe and barley in your gravelly grounds, when you fow them broad-caft and harrow them in, for thefe foils have often, by the ill management of farmers, been the ruin of many crops of thefe grains, by the binding quality of the gravel, when great rains have fell on them prefently after the feed has been fown and harrowed in; which misfortune to prevent, the judicious fort will plough, fow, and harrow in their peafe, or barley, every day, as being a much furer way to be free of fuch a cafual lofs, than to plough on and not fow till a whole field of feveral acres is ready and is fown all in one day; becaufe, by daily fowing of the pea-feed, the hazard is not near fo much, as when feveral acres are fown and harrowed in at once; and if the feed is bound in by the running like a pancake of the furface, when wafhed by great rains, the peafe and barley fprouts are

kept

kept in, and often caused to rot, because they cannot make their way out through the hard, crusty top earth of gravelly grounds. To prove this, I shall recite a case as it happened to a farmer at *Chessum* in *Bucks*, where having sowed his seed broadcast in such a soil, and harrowed them, great rains fell presently after, and bound them in so tight, that there was little hopes their sprouts would ever have gotten through the top earth. But, as it luckily happened for this farmer, a neighbour's team being at harrow, in the next field, run away with the harrows, and got into this man's field that had been sown with pease; and it was plainly observed, that, where the harrows were drawn, there was a second crop of pease; but, where they missed, there was none. Hence it is, that several farmers take the caution, to give their pea-crop, sown in the broad-land fashion, a harrowing just as their heads are peeping out of the ground, which not only loosens the earth, but moulds it up to the pea-heads, and oftentimes greatly improves the crop.

It was at *Michaelmas*-time, that a farmer, at *Dagnal*, in *Bedfordshire*, within a mile and a half of my house, took possession of a farm, and being a new tenant, with a great deal of business to do at entrance, he had not time to plough up a clover-lay that he came to, and therefore could not get it ready for sowing wheat in the same; and,

Chap. I. PLOUGHING FOR PEASE. 359

and, when afterwards he had more leizure, he thought the ground would be too four, and the wheat expofed to the fpoil of birds and frofts, to come to profit; on this he came to a refolution of giving the clover-lay one ploughing in broad-lands, and let it lie all the winter for the turf and roots of the clover to rot, till *February* following, when he harrowed the furface plain; and, after it had lain fo about a fortnight, he fowed horn-fhavings all over the ground, and ploughed them in crofs the laft way of ploughing; then he immediately fowed it with peafe, and it proved a vaft crop; for this grain in particular, loves both a hollow earth and a hollow dreffing.

Here were two faults attended thefe crops of peas: one was, that the farmers did not keep their ground in fufficient ftrength of heart; if they had, the richnefs of the foil, very probably, would have forced the peas into fuch a quick growth, as to have caufed them to grow fafter than the flug could have eaten them. The other was the farmer's rolling his peafe in the day-time; which was wrong acting, becaufe then the flugs were moft of them retired into their fubterranean cells, and lay fafe out of the fqueezing of a roll. Therefore he fhould have obliged his fervant to roll his pea-crop about one or two o'clock in a morning, when the flugs were moſt buſy in feeding on it; then he might expect to have had all, or moſt of them, ſqueezed to death, by his eight-feet-long heavy wooden horſe-roll, and have preſerved his peas.

CHAP. II.
OF THE SORT OF PEASE.

MANY are the different ſorts of peas that are ſown in this kingdom, whereby the farmer has his choice for a proper ſoil, clime, and ſeaſon; but for my part, I ſow at preſent only three ſorts, that I have found to anſwer beſt; and they are the horn-grey, maple, and blue-pea. The firſt is accounted, with a great deal of reaſon, the hardieſt of hog-peas, and is of a good ſize, and preferred to all others for the chalky ſoils, where it will admit of being ſowed forwarder than any other, which is a perfection very much in favour of this ground; becauſe by this means it enjoys all the ſpring wets in this dry ſoil, and thereby obtains an early head that covers, ſhades and defends the earth about them from the ſcorching heats of the ſummer, that otherwiſe would dry up and wither their roots: alſo on gravels this pea has the like ſucceſs, which makes the farmer ſow them in theſe two

ſoils

soils presently after *Christmas*, where they strain them in after the plough in broad-lands, and harrow them afterwards, till they are almost ready to appear; insomuch that about *Ivingboe*, and many other places, they sow no other sort but this in their chalks and gravels, on one ploughing after wheat or barley crops.

The maple is a larger and sweeter pea for the hog, but somewhat more tender: this is chiefly sown all over the *Chiltern* country in *March*, and is found to make the best returns of all others in both grain and straw; and therefore the present practice mostly runs in an equal mixture in these two sorts of peas, that are sown together about the seventh of *March*, in several ways and in several soils, and this for the better assurance of a crop; for if the maple should miss, it is hoped the horn-grey will hit; which is, having two strings to the bow; as is also often done after the very same method, by sowing a mixture of seven pecks of peas, and seven pecks of beans on an acre: this way proved exceeding well this spring, when the beans were the best crop ever known; while many of the peas were spoiled by the chill of wets and frosts that attended the whole month of *May* 1732.

The blue pea is a more tender sort, and is sown both earlier and later in *April*, on prepared ground that has been twice ploughed: this pea

oftentimes

oftentimes returns great crops, is a most sweet sort for any of the culinary uses, and also fats swine in a very short time; but the haulm or straw of this pea is not so good for horses, as that of the horn-grey or maple; because this is apt to gripe them, and bring on the yellows, that sometimes prove fatal if not taken in time; and is first discovered by the white of the eyes being turned yellowish; for which, bleeding, and cordial, sweetning drink is the best remedy. This misfortune is said not to be owing so much to the straw, as the peas, that are generally left in after thrashing, which being of a softer nature than all other peas, is thereby more endowed with this pernicious quality. But there is a great difference in the nature of this pea; some of them will boil in a little time to a great softness, others require a longer time, and some will never boil so sweet and kind: the reason of this I take to be owing to the particular juices of the earth wherein they grow; for that of the stiff clays is commonly churlish and sour; that of the loams much better; but that of the chalks, sands, and gravels exceed all others, in sweetness and quality; therefore the best boiling peas are allowed to come off the three last soils.

The horn-grey is best to grow with beans, because it will endure to be sown with them at *Candlemas,* agrees with all sorts of ground, and will better bear with one ploughing than any other: thefe two kinds have each their reciprocal benefits; for the pea will fhade the roots of the bean, and keep off the parching heats and droughts of the fun and wind. The bean alfo will, by the ftrength of its upright ftalk, keep up the weak, bending haulm of the pea, and caufe it to kern and ripen kinder and fooner.

The vale-men in their open, low fields, whofe ground is in heart, will not fow peas alone, or beans or peas together (which they call half-ware) becaufe then the fheep cannot weed amongft them, and eat up the wild oat and cur-lock which often infeft them; but in the clean beans they turn in their fheep till bloffom-time to eat up thefe weeds; then they take them out, left they rub them off.

This I was an eye-witnefs of. The father-in-law fowed his inclofed, gravelly, loamy field broad-caft with hog-peafe, and ploughed them in with the pecked fhare two-wheel plough; which, becaufe he ploughed the ground but once, and a bafhing wet time fucceeding, bound the earth and buried the peafe. But the fon-in-law acted wifer, for in the fame foil, in a field near his father's, he like-wife fowed his hog-peafe in the fame manner, but ploughed them in with a foot broad-fhared plough, which covered them fo fhallow, as only

juft

just to cover them. Both which ways were intended to shelter the pease from too much drought. The result was, that the father-in-law lost his crop, but the son-in-law had a full one, though both were sown much about the same time.

There was a gravelly, loamy, inclosed field, sown in a random way with the maple-pea, about the twelfth of *April*, and harrowed in on only one ploughing, and, a kind season following, three acres of them yielded near an hundred bushels. But such late sowing of maple-pease is too great a hazard for a prudent farmer to run, for those sorts should never be sown later than *March*, lest they run into straw, and not corn; as it happened in another case at *Wardscomb*, near *Ivinghoe*, where a farmer sowed some grey horse-pease on the thirtieth of *April*, and had only straw in return. Had he sown blue pease in their room, or the *Essex Kodeing* white-pea, or beau-dye, or some others, it is very likely he might have had a good crop. The puffin, the kid-pea, &c. are tender ones, and so tender, that, if they go once away, in clay-ground, by cold weather, they never recover. But the horn grey is so hardy as to come and go (as we call it) several times, and yet be a good crop at last.

Chap. II. OF PEASE.

April gives many of the *Chiltern* farmers an opportunity to improve their land, by sowing it at this time, with a sort of pease proper for the season of the year; for on this very much depends the success of a pea-crop; a truth often proved by the fatal effect of a wrong conduct, when it is employed in sowing that sort of pease in *January*, *February*, or *March*, which should not be sown before *April*, because their tender nature renders them unable to withstand the severe assaults of northerly and easterly winds, which in these months generally blow in a more nipping and sharper degree than at any other time of the year. From hence it is, that, though the soil is duly prepared, and the seed got in perfection, and the sowing of it performed in the best manner possible, yet all this will not secure a plentiful crop of pease, if the farmer sows a wrong sort at an improper time. After many years experience, we *Hertfordshire* farmers have found none to answer our interest so well as the horn-grey hog-pea, the maple, the blue, and the white. It is these four sorts that we chiefly sow in our inclosed fields, in refusal of all others, because we find, by repeated experience, that these best agree with our various soils, when sown, in a proper manner, at the right time of the year. If we should sow a horn-grey pea in *April*, the farmer would deceive himself

in

in his expectation of a corn-crop, because, instead of that, he would get little else but straw; for this pea is of so hard a nature, and so slow of growth, that it requires six or seven months to ripen into a plentiful crop of pease, when a blue pea, and others as tender as that, will be in perfection for mowing, or hooking, in four months; so different are the qualities of these two sorts of pease, though both grow the same year in the same sort of ground. It is, therefore, that, if a blue or other tender pea was to be sown in the month of *January* or *February*, the crop would consequently be lost by the severity of the weather, because the *Hampshire* kid-pea, or kidwell, the maple, the *Essex* roading, the puffin, the *Spanish* mulatto, the rouncival, *Dutch* admiral, the marrow-fat, the non-pareil, the blue union, &c. are all of them of so tender a nature, that, if they are thoroughly pinched and nipped by a very hard frost at root and stalk, they can never recover a right growth again, but will shew their sickness and decay by their red and withering heads. At best, a pea-crop is subject to more fatal accidents than any other corn we sow besides; insomuch, that, in the random, common, broad-cast way of sowing pea-seed in fields, we hardly get a full crop of them above once in three years, because the cold, the drought, and the slug are all enemies

mies to pea-crops, when they happen to be in too great a power for their regular growth; which leads me to write on their particular damage, for the knowledge of the difeafe is fometimes half the cure.

About *Rickmanfworth* in *Hertfordſhire*, they fow a forward hog-pea, called here the puffin-pea; which being fit to cut in *July*, or the beginning of *Auguſt*, they commonly plough and harrow in cole feed on the fame land, for their fuckling ewes to feed on them the winter, or fpring following; that by fuch juicy food they may expeditiouſly fat their houfe-lambs for an early market at *London*. Now the great *Spaniſh* murotto-pea, that was fown in drills at three feet and a half diftance, after being well houghed and feveral times turned in the field, is in hot fummers ripe for feed about the beginning of *Auguſt*; and fo are the *Carolina*, *Barns*'s mafters, marrow-fats, and other forward forts, that were fown in drills out of the hopper of the three-wheel plough, or out of a hand-box; both which are much truer ways than out of the naked hand: but the drill plough I muſt here once more recommend, as the beſt inſtrument that ever was invented for tenants to make uſe of, for getting a livelihood and paying their rents; yet, where this could not conveniently be had, I have feen them make a drill

a drill with the two-wheel turnrife-plough, whose chizzelly point is made from one inch and a half to two inches breadth, and then follows a man with a long box in his hand, out of which the peafe run leifurely and gradually into the drill; and when all the field is fown, they harrow long and crofs-ways. At another place, they fow their grey and yellow forts of peafe in drills, at twenty inches afunder, in fields, and afterwards employ the horfe-break between the rows, and then immediately ufe the hand-hough to rake and hough the earth up to the roots of the peafe; and this they do twice or thrice in a fummer, befides turning the rows to the fun now and then for their better ripening. The *Cobham* grey is the lateft ripe. If the poor-land white pea is fown three years on the fame ground, it lofes its colour. With us in *Hertfordshire*, we fow the horn, and *Windfor* grey hog-peafe, that are early fown, and late ripe. The maple-pea, for either hog or boiling, we fow in *March*, and it is ripe in *Auguft*. The *Hampshire* kid and the beau-dye are alfo fown for either ufe; but the blue pea is a very convenient fort for our farmers, becaufe we fow them late in *April*, and cut them in *July*, or *Auguft*. The *Cobham* grey is a large, fattening hog-pea, generally ripe with wheat, but is apt to fhed, if they ftand a little too long before they are hooked. In *Effex*, about
Chelmsford,

Chap. II.   OF PEASE.   369

Chelmsford, they fow the *Cobham* pea broad-caft in their clay land, and the yellow pea in their light.

The common and rouncival maple pea, may be called a hog-pea, or a boiling pea: In *Hertfordshire* we fow it as a hog-pea, but it is made ufe of in many places as a boiling pea; and where they do this, they call it a grey-pea, as in *London,* where the women boil it, and cry it about the ftreets for grey-peafe. It is of a good fize, and a tender fort, and therefore we *Hertfordshire* farmers, that know its nature, dare not fow it till the latter part of *March,* left the frofts take it, and the crop be fpoiled. For this reafon it is, that the maple-pea is not fit to be fown in *February,* and they that fow in that month, or in the beginning of *March,* run the greater rifque of the crop's being damaged by the cold weather that generally blows at that time; but, when it is fown on a pretty loofe loamy earth about *Lady-Day,* we find this very pea to yield commonly the beft crop of all others fown in the random way; and for this reafon it is fown in the weftern part of *Hertfordshire* more common than any other pea whatfoever, as being after many trials found to be the beft common hog-pea for this country. But there is a rouncival fort of the fame nature, that requires much the fame culture.

CHAP. III.

OF THE TIME OF SOWING.

IN this chapter, I shall shew the skill of some farmers, as it relates to their sowing of hog-pease in *January*; which, to do, I am to observe, that the pease, sowed in this month, should be the most hardy sort of all others, and the soil a very dry one, because the severe frosts, that generally happen now and hereafter, would kill a tender pea sowed in wettish or cold earth. But when the hardy horn-grey hog pea is sown in a chalky soil in *January*, notwithstanding the cold and wet weather that may follow, and cause their heads to look reddish and pine, yet this good property attends them in such a earth, that they will recover again as soon as the weather becomes calmer and warmer. Therefore in such a chalky soil, whether the preceding crop was wheat, or barley, it matters not; the farmer here sows his pease either broad-cast, to the quantity of three or four bushels on an acre, and ploughs them in, with the foot or wheel-plough, which way saves the charge of a sower: or else a man sprains the pease in out of his hand, in a thorough or furrow made by the plough, and then the plough turns the next thorough, or furrow of earth upon them, and so on, throughout the field.

Many

Chap. III. OF SOWING.

Many crops of peafe have been loft, by sowing their feed too early in wettifh, or clayey loams; for, in this cafe, the foil and feed ought to be duly confidered in their nature. If a puffin-pea, a *Hampfhire* kid-pea, a kidwel-pea, a maple-pea, or a blue-pea, fhould be fown in a field, in thefe foils, in *January*, or the beginning of this month, it would greatly endanger their crop, by chilling and killing them in their infant growth, if much frofty and wettifh weather happens in that time, when a horn-grey, *Windfor*, iron-grey, and fome others would ftand and efcape the danger.

This moft hardy pea is often experienced to grow and thrive in chalky, gravelly, and other dry land, and alfo in clays and ftiff loams, if fown early enough in them; and, as it is, for thefe reafons, the moft general pea, of all others, for fuiting moft forts of foils, and bearing to be fown the earlieft of any pea of the hog-kind, it is preferred, for fowing in vale, and moft *Chiltern* grounds, to all others. It is this excellent pea, that will grow in mixtures with thetches, or with oats; but beft of all, with the horfe-bean, fown with it, in the month of *December*, *January*, or *February*, and is annually found to produce, in this manner, very great crops. And, although the beans afterwards cannot well be feparated from the peafe, they are little or nothing the worfe for fatting swine, or feeding horses; at least, we *Hertfordshire* farmers make no objection against the mixture, for, the species being inseparable, it is often known, by woeful experience, that pease are subject to be killed by cold, wets, hail, &c. even till *June:* but, of all pease, none will stand the severity of weather like this, for this has been found to go, and come again, several times: that is, although it seems almost killed by frosts, and chills of wets, yet, if any pea will recover it, this will, and often does, when others die.

Of all the tribe of pease, this, the *Windsor* grey-hog-pea, and the iron-grey pea are the firmest and hardest; insomuch that is a rule with the judicious farmers, to give the horn-grey, or *Windsor*-grey, or iron-grey pease, as the first sort, to the swine, that are to be fatted with pease; and the maple-pea, blue-pea, or the white or other softer pease, after they have sufficed their keen appetites, and got some fat on them; for, if they were to be fed with the maple, blue, or white pea first, they would suffer in their flesh very much, before they would eat them. The horn-grey pea is bigger than the *Windsor*-grey pea, or the *Essex Roading* pea, and serves the farmer as the best sort of pease for manger-meat, when they are dry enough to give them with safety; and, when they are otherwise, it is our
common

common practice to carry a parcel of them to the malt-kiln to be dried and split, and then with a mixture of oats and chaff, they become excellent corn provender, and maintain our cattle well under the hardest labour. And such stress is laid on this horse-food, that there are few farmers, in our *Chiltern* country, but what prefer these pease to beans, and all other corn, because the pea feeds and fattens, and yet does not heat the blood of the beast, like the bean, which, by some, is, on this account, refused as a manger-meat, as tending to the breed of farcies, &c. especially in coach-horses, that are obliged to run and heat themselves more than cart-horses. Some, therefore, will sell their oat-crop, and keep the horn *Windsor*, or iron-grey pea, to supply their place in the manger, because pease will go further than oats, in feeding horses.

In the parish of *Ivinghoe*, in *Buckinghamshire*, a great deal of their land consists of a stiffish chalk, of a nature between a dry hurlucky, or what we call a short sugar-plum chalk, and a white clay. In this soil, a quaker farmer was ploughing and sowing the horn-grey hog-pea, in *Christmas* holidays, because this horn-grey pea is of a pretty large size, but not so large as to deserve the name of a rouncival pea, and is, past contradiction, the hardiest pea of all others whatsoever; for this pea will stand the severity of the frosty weather, when all others are destroyed by its violence, having this particular beneficial quality belonging to it: that though through the vehemency of the cold, chill air, the infant heads and stalks of this pea are so pinched by it, as to turn even almost as red as a fox's tail, yet, on a succession of mild weather, they will recover their pristine verdure, and flourish again; and this they will do several times, as frosts and mild weather alternately happen; a character I can give no other pea, with such a certainty, as I can of this. Yet as hardy as this pea is, when it is sown in clays, or stiff loamy ground, it has suffered by long frosts, and chilly wets, which have overpowered its hardy quality.

In very dry summers, especially in blossoming season, when pea-crops in such soils and weather are not able to perfect their blossoms for want of moisture, and then the crop is, in part, or fully lost. A case that often happens in chalky soils, not above three miles distance from my house, as I have heretofore observed, where several of their farmers are weary of sowing peafe, at any time in this earth, because of this damage that makes them suffer more or less in very dry summers; and, therefore, in despair of better success, they have sown oats, where peafe, by the course of sowing, should have been the crop; so powerful are the effects of very dry weather on pea-crops

crops sown in these soils, that, though the hardy horn-grey pea is sown in them quickly after *Christmas*, in order for their obtaining an early sufficient cover to shade their roots, yet even such early sowing is not, in such dry hot summers, capable of preventing this misfortune; much less when pea-seed is sown later than ordinary in such loose dry earths.

CHAP. IV.

OF THE QUANTITY OF SEED.

THREE bushels sow an acre of pease in the random way, and a bushel and a half in the drilling way, or less, if they are set with a stick in rows; but the setting of pease in fields is now in a manner left off every where, since the drilling of them is found to be a much cheaper, quicker, and far better way.

There must be four, or, better, five bushels of horn-grey seed sown out of a man's hand in all the thoroughs that a plough makes in one acre; for here he is obliged to sow the pease so thick, that they may come up thick and close: this prevents both frosts and droughts entering the ground to hurt their root; and, the older they grow, their stalks will the better answer this beneficial purpose, and secure them betimes

against

against the sharp winds of *March*, which are very apt to dry and blow away this loose earth from the roots of all corn that grows therein, and leave many of them so bare, as to kill them.

When we sow horn-grey pease alone, in the broad-cast method, in broad-lands, we commonly sow in a poor soil four bushels and a half, or five bushels on an acre; because, in this quantity, we allow some for loss by birds, mice, slugs, and other destructive incidents: but when a man sows these pease by spraining them in, out of a man's hand, after the plough, less than four bushels are sufficient: three, or three and a half, are the common allowance in this way of sowing; for this spraining mode saves seed, whether it is done in ridges, or broad-lands.

CHAP. V.

OF HARVESTING PEASE.

IN 1740, the field hog-pease ran so fast into pods, that the bloom was hardly perceived; and, when it does so, we say the pea steals a bloom, and then we reckon it a sure sign of a plentiful crop, which accordingly happened, for they corned extraordinary well in most places. We had the puffin, *Windsor*, and horn grey-pea, poplar, and maple hog-pease ripe,

Chap. V. OF HARVESTING PEASE. 377

ripe, which are cut feveral ways. In *Hertfordſhire* we do it two ways; one is by the pea-hook, which has a five feet long wooden handle, with a cutting-iron made a little circular at its end, about a foot in length, and an inch and a half broad. The other is by the ſcythe. If the peaſe are very thick and long, we hook them; if thin, we mow them. When they are ſown in two bout-lands, we commonly hook them and not mow them, becauſe it is difficult to mow them, while they are in this poſture. In *Kent* they make uſe of two inſtruments in this work, called hook and hinks, or hook and ſwipe, which their men dexterouſly manage; and, when all is cut down and dried, they make bands of the fame, and bind the peaſe up in bundles for carrying home.

In caſe great rains fall after cutting, the wads muſt be turned now and then, and thereby you will prevent the opening of the pods and ſhedding of the peaſe in a great meaſure; and this is ſo well obſerved by careful farmers, that even in wet time, and when the rains continue long, they will turn them, becauſe it will keep the very undermoſt peaſe from opening.

It is the opinion of a certain farmer, that when peas have been ſo wetted in the field as to ſprout, if they be dried afterwards by fine weather,

ther, they are the sweeter, and that he has observed his hogs eat them better, and thrive faster, than when they have eaten them very dry and hard. But then such pease must not be inned too damp; if they are, they will mowburn, or turn mouldy, rotten, and stink.

BOOK VIII.

Of the CULTURE *of* BEANS.

CHAPTER I.

OF PLOUGHING FOR BEANS.

THE vale ancient and common way of getting a crop of horse-beans, is to plough the ground but once for this purpose; and that is generally done in *February*, after a wheat or barley crop had last grown, when they sow the same land with bean-seed twice in a place, broadcast, to the quantity of two bushels on each half acre ridge-land, and they then immediately, with the foot-plough, plough in the seed, by casting down the earth; for it is very rare, that any plough their ground at this time by ridging it up in these parts. When the sowing and ploughing part is finished, they draw the harrows over all the land; but some neglect harrowing, till the beans are ready to peep out of the earth, the better to give them an easy passage. And thus, in a favourable summer, they enjoy the best of horse-bean crops; for no land exceeds this vale sort of stiff black earth, for profitable bean-crops; which is the chiefest soil of the vale of *Aylesbury*.

The method of sowing beans is so material an article, that I thought it absolutely necessary to be particular in my remarks on it, because on it depends very much the good or bad success of the crop. On this account it is, that some of the brightest farmers that I know, who rent several inclosed farms in the *Chiltern* country, whose land consists of gravelly loams, chalky loams, and intire dryish loams, believe, that all such dryish soils require to be sown with a larger quantity of seed, than either a clayey soil or a wettish loam: because, as the horse-bean affects a stiff moist earth, wherein it may lie wrapped up in a close coverture of the same, the less will serve an acre, for here the bean receives a nourishment on all the sides of it in a constant supply; for, let the weather come how it will after the feed has been thoroughly soaked by rains, such stiff earth will communicate some degree of moisture for carrying forward the after-growth of the bean-crop, while, in a poor, dryish, loose soil, the seed-beans lie in a hollow loose condition

tion or coverture, and are thereby subject to suffer by droughts of little duration. Hence it is commonly said, that some dry earths can easily dispense every summer's-day with a shower of rain. What then must be the effect of long droughts to bean-crops, where the seed was sown thin? Why, a short dwindling stalk, a few blossoms, fewer beans, and a great many weeds. Now, to prevent these fatal misfortunes, the farmers, I have before mentioned, sow five bushels of horse-bean seed always on each acre of their land, either broad-cast and plough them in, or a seeds-man follows the narrow two-wheel plough, and strains or sprains in the beans; that is, by tossing or throwing the seed in a direct line out of his hand all along the furrow the plough makes and leaves, which is covered by turning down the earth of the next furrow, and so on till all the field is sowed; and none have better crops in the *Chiltern* country than these farmers. The benefit then of this management is this, that, as the beans are sown thus very thick in a dry loose earth, their stalks will grow up so close together as to shade each others roots, and, by this means, get a good cover in a little time, against the parching heats of summer-weather, and thereby the moisture of rains and dews will be retained a long time after falling,

to

to the great advantage of such crops; as has been often proved in dryish summers, in the aforesaid dryish earths, of loamy gravels, loamy chalks, loamy sands, and in intire loams of the *Chiltern* country; which also keeps their ground clear of weeds, in a hollow condition, and from being exhausted, in a great measure, of its fertile quality, by the attraction of the powerful sunbeams, which in earths, where the seeds of corn are sown too thin, is often the case, especially in dry hot summers.

A fourth way to improve a crop of horsebeans in the *Chiltern* country, as I have seen it performed between *Watford* and *Hempstead*, is: they set them in a gravelly, loamy, inclosed field, that was ploughed hollow, by the short dibber, made with the upper-part of the handle of a shovel or spade, cut off within five or six inches of the hollow handle into a small round point, which is shod with iron about three inches in length; this women job into the ground, and then immediately drop into the hole a horsebean, and so they proceed in a very quick manner, making the holes by a line, perhaps, fifteen or twenty yards long; and, when one row is finished, the line is moved at a foot or eighteen inches distance, and so they go on throughout a field, till the whole is compleated. Thus the

women

women work for nine pence, setting every peck of horse-bean seed, which they carry in an apron before them, and with great agility take out and set by this dibber; the latter in the right hand and the other in the left, at three inches asunder each hole. Afterwards they harrow all the ground, and, when the beans have got a few inches above the earth, they hand-hoe them.

About *Swanburne* in *Bucks*, their ground is so heavy a clay, that they commonly let it lie three or four days after ploughing, or more, before they sow in order for the weather to slacken and shorten the surface, and then they are obliged to sow their horse-bean seed, broad-cast, and harrow it in; for, if they plough it in, they say they shall have no beans, contrary to most other parts of the same vale, where their land is of a blackish, stiff loam; and, therefore, they are forced here, to plough in their bean-feed, to have a great crop, for a small frost presently shoals and crumbles this stiff, fat, black earth; yet many of the beanheads, in this soil, will get doubled, in making their way out of it; and therefore they forbear using their harrows for several weeks after sowing the beans, because then they relieve these doubled heads, by the scratching of the tines, which loosens the earth, and lets them easily out.

This sort of ground, as it generally lies flat, low, and wettish, they plough but once for sow-

ing

ing it, either with peafe or beans in the fpring time. And why they do fo is, becaufe, if they winter-plough for thefe, they commonly fuffer, by the lofs of great part of their expected good crops; by reafon if they break up their heavy ground in any part of the winter, it is the more expofed to the flug and wafh of rains; and to the lofs of its fpirits or ftrength. Whereas, when their ridge-acre, half-acre, or rood-lands, lie all the winter undifturbed till *February* or *March* in whole ground; then their earth will plough up in a frefh, lively, dryifh condition, and revive and nourifh the pea, or horfe-bean feed in fuch a manner, as to carry the crop forward in a vigorous growth: and efpecially, if they enjoy a right feafon at fowing-time. This item, againft ploughing up ground in a wrong feafon, is not only neceffary to be obferved by vale-farmers, but alfo by all others; for a man may beggar his lands by many ways of ill-managing it; and, in particular, by this of ploughing it too often, or too foon; which further fhews, how much the art of farming depends on the conduct of the mafter, or his ploughman. But fome may think, that, by giving thefe vale-lands only one ploughing or fowing-time, the earth will be too rough and clotty for the feed-beans. On the contrary, the vale-farmers like it beft, when it is fo; for then they know a heavy thorough, or furrow,

will

will be turned on the horse-beans, lie close and moist on them, and shelter their roots afterwards from the violence of frosts and droughts; and their heads, or first sprouts, will easily make their way through their thick surface; because, as these vale-earths generally partake of a marly nature, their upper part is presently shattered by a small frost into a crumbling loose condition.

CHAP. II.

OF THE SOIL FOR BEANS.

LAND cannot well be too rich for a bean-crop, because the large high stalks, and great numbers of beans growing on them, employ so much of the goodness of the earth in their nourishment, as makes it necessary beans should be sown in a rich and not poor ground; insomuch that, if the ground is not in good heart, on which beans are sown or set, there is little likelihood of their being a good crop; for, if a rich soil, in a dry summer, returns but a poor crop of beans, what can be expected from a poor soil? Therefore, all vale and *Chiltern* clays, and stiff loams, are certainly the most proper sorts for beans; and, as stiff earths are generally the rankest or richest sort of land, I have seen in such ground, in the vale of *Ayles-bury*,

bury, many crops of beans, that have been equal in value to the like quantity of ground sown with wheat or barley, seventy, eighty, or more pods of horse-beans growing upon one single stalk ; and then it is, the vale-farmer obtains such a bulky crop, as obliges him to make his huge ridge-stacks, or cocks of beans abroad. Then, if he be a rich man, and able to bear stock ; or, if he is poor, and his landlord rich, and willing to trust him with two or three years rent, such a tenant stands a rare chance of selling his old beans to a great advantage ; for it is generally observed, that horse-beans very rarely hit three summers together.

A poor dryish land may be made capable of producing large bean-crops ; which is an article that more than ordinary deserves attention, because it is of great importance to the lands of owners of some sorts of dryish land, which they think cannot bear a bean-crop, by reason of such its dryish nature ; which the generality of people allow to be a quality repugnant to the profitable growth of a horse-bean. Yet, for all this, thousands of acres of land may be sown, or set, with horse-beans, to advantage, that never were. A sandy soil may bear a plentiful crop of beans : but, when I mention sand, I would be understood to mean such a red, white, or yellow lean, dry sand, as many places afford ;

no,

no, it is a blackish, rich, dry sand, or a springy or other wettish sand, no matter of what colour, that I mean here, which is capable of being improved by a bean-crop.

Land, fit for wheat, is commonly fit for horsebeans, and here they are often sown on purpose for a change of seed to the ground, as well as for profit otherwise.

CHAP. III.

OF THE SORTS OF BEANS.

TICK-BEANS may be justly named *the large horse-beans*, because they are chiefly sown or set for the use of horses, as well as the small field horse-bean; and this more of late than ever; for that these, being near as large as the small garden hotspur bean, they turn to the best account for splitting in a mill, to give with chaff or bran, or oats, as manger-meat to horses; and also for their great bearing quality, which in rich ground, in a kind season, and under good management, generally produce vast crops. Both the tick horse-beans, and the common small horse-beans, are set alike in some parts, but in a different manner in others.

CHAP. IV.

OF SETTING BEANS.

ABOUT *Harrow, Hoxton, Acton, &c.* in *Middlesex*, in several of their stiff loamy fields, they, with their large swing-plough, plow three four-bout-lands into two, and harrow them almost plain; then, by lines laid over four or five of these broadish lands a-cross, women set horse-beans in holes, which they make along each line at two inches asunder; and so proceed till the whole field is finished, and then they harrow it all over once in a place. On the 9th day of *February* 1742, as I rode along, I saw great numbers of women at this work in these parts. When the horse-beans are at a proper height, they hoe them with hand-hoes once, and sometimes twice; but there is another way made use of here by some of their farmers, where their ground is more a clay; and that is by ploughing up their wheat-stubble four-bout lands about *Christmas*, and letting them lie till this month, when they sow their horse-beans, broad-cast, and harrow them in; for, if they should sow first, and then plough them in, they would be so buried, as never to become a tolerable crop. And so they do sometimes in three-bout lands, where the ground lies wet;

Chap. IV. OF SETTING BEANS.

and when they have sown their horse-beans, broad-cast, over these narrow lands, they strike up their thoroughs with their swing-plough, and afterwards pull up their curlock-weed by the hand, but chop up the couch-grass.

About *Gainsford* and *Rouslup* in *Middlesex*, their land lies so low, that they are obliged to keep it up in high ridges, to preserve the corn dry, and prevent the damage of waters. On this account it is, that they set their horse-beans by the hollow-handle dibble, and by a line laid a-cross the ridge-land, and so another cross row, at about a foot distance, thus leaving room for men to hand-hoe afterwards between the rows; and, when the whole field is about four inches high, men begin to hoe longways between the rows of the ridge-land; and about five weeks after, they hoe a second time; and, at this last hoeing, they pull the earth upon the roots of the beans; which operation not only kills the new-sprouted weeds, but supports the stalks, keeps them shaded at the bottom from the droughts, and adds a new fertility to their growth. Two bushels of horse-bean-seed set one acre in this manner.

Sometimes they raise a good depth of mould by the plough, on a shallow surface, for the better setting and growing of a crop of horse-beans, which is performed in land of a stiffish, wettish

wettish nature, situated near *West-Hyde*, in *Rickmansworth* parish, that is not so stiff, nor lies so wet, but that they plough and sow in broad-lands, on a wheat-stubble, particularly for a bean-crop, when they plough and sow it after the following manner: They get ready two teams, one draws a swing-plough, with three horses in length; the other draws a swing-plough, with five or six horses in length; and this on purpose to make a shallow soil a deep one: For the first three-horse team turns up a shallow earth; the other team follows immediately after, and turns a deeper earth on that. When this is done throughout the field, they give it one harrowing, and then begin to set their horse-beans by line.

CHAP. V.

OF HOEING BEANS.

THE benefit of hoeing was more apparently seen this year than in many others; because the long, cold, dry spring, and dry summer, 1740, caused the blossoms to dry and fall off for want of sufficient moisture, and even killed many a bean-stalk after it had got pods on. About *Pinner*, near *Harrow*, in *Middlesex*, they were more than ordinarily sensible of this,

Those

Chap. V. OF TOPPING BEANS. 391

Those horse-beans that were drilled and hoed, were good crops in general; but those that were ploughed in, or harrowed in, were for the most part as bad. Here they sowed their horse-beans in drills, at two feet asunder, which was hoed twice in all, for four shillings an acre; and one man in this work would hoe over an acre in one day, for which he had two shillings; and the same when he hoed them a second time: However, by one fault is known how to prevent another; and they are now resolved to sow all their beans in drills for the future.

CHAP. VI.

OF TOPPING BEANS.

AS soon as ever the blossom falls off from the bottom of the broad bean stalk; that is, as soon as the lowermost kid appears, then cut or pinch off just the head of the stalk, for then as the stalk kids upwards, this method will stop its shooting in length, make it corn better, and less liable to the dolphin-fly; but this work is rather too tedious to be done among horse-beans.

CHAP. VII.

OF ROLLING BEAN-GROUND.

IN the *Chiltern* country we commonly roll those beans that are to be mowed; some as soon as they are ploughed in, others as soon as their heads are all out of the earth, to close the ground about them, and better secure them from the damage of droughts, and make it lie even, that the scythe may work the better. Some harrow them presently after their appearance, saying it loosens the earth, lets in rains, prevents the growth of weeds, and, if the tines split any of their heads, they will spread and grow into the more stalks.

CHAP. VIII.

OF THE FLY, &c.

THESE insects always begin to make their lodgment on the top of this vegetable, and increase downwards till they kill all or most of the growing beans; therefore, when they have first got possession, mow off the heads of the beans with a scythe, and the fly will never rise again, for they cannot get upwards. The next

Chap. VIII. OF THE FLY, &c.

next way is to do it by turkeys: A certain farmer's wife used to scold at her servants for letting the turkies go into a field of beans near her house; but instead of mischief, they did great service; for they proved excellent weeders, by picking off the flies for their food, and caused more beans to grow on four acres, than was in forty of her neighbour's that summer.

A field had been a meadow time out of mind, in the parish of *Studham* in *Hertfordshire*; but, the grass decaying on it, the owner, to recruit it, dunged it well all over, on which ensued such a dry summer as burnt up the crop. The next year he sowed *London* coal-ashes over it, and had a tolerable return; however, having seldom or never had a full crop of grass on this meadow-ground, he resolved, as soon as a composition was made with the tytheman of the parish (which was then about to be done) to plow it for sowing corn; accordingly the tythe was agreed to be paid in money, and then he up with this meadow, and sowed it with beans, which were so destroyed by the grub and cankerworm, that at harvest there were hardly any beans; for these insects, having had a long series of years to increase their breed undisturbed, multiplied prodigiously. After this he was going to sow it with wheat only on one plowing, but

but was dissuaded from it, and instead thereof ploughed it up and harrowed it several times, till he got it fine enough to sow it with turneps, which he did; and, after they were eaten off, he gave the same ground one plowing, and harrowed in barley, and after that wheat, and had excellent crops, free from any damage by the canker-worm; which by this time were most, or all, destroyed by the plough-share, hoe, and the tines of harrows.

CHAP. IX.

OF THE PRODUCT OF BEANS.

IN a gravelly loam in the *Chiltern*, a wet summer has produced 260 pods on six stalks, that grew from one root; and upon one single stalk ninety pods have been found: And from one acre forty bushels of horse-beans have been got, which makes some of opinion, that a full crop of these will pay a farmer as well as a crop of wheat, especially where they dung for beans, and immediately after sow wheat on the same without dressing, as many do in some *Chiltern* countries, to their great improvement. My neighbour having a field of three acres to sow, whose soil was a flat loamy

loamy earth, he sowed five bushels of beans amongst pease and thetches in a very dry season. The pease and the thetches missed, but the beans that grew very thin stood it, and became so well corned, that he had an hundred bushels of thrashed horse-beans off these three acres. Another *Chiltern* farmer had five quarters of beans and pease that grew together off one acre, which is a good Vale-crop.

CHAP. X.

OF HARVESTING BEANS.

IN the vale of *Aylesbury*, where the best of bean-ground is, they mow all their beans with the bare scythe, in *swarth*, as they call it; that is, they mow their beans towards the beans, and each mower has a boy to follow him with a fork to lay them in wads, in which posture they let them lie to wilk and wither. Next, they lay two of these wads in one along the ridges of their land, and directly draft, rake, and leafe all with their own folks, I mean those of their own family. When dried, they carry, draft, rake, and leafe all their bean land over again. But, in the dry summer 1740, they mowed their beans out in swarth, because they were

were so short and thin, that they might easily do it, this way; for, when they mow beans in swarth, it is, because they are a large crop, and stand leaning inwards, which they are obliged to observe; else when they mow beans against their bending, they call it *Throating*, that is, moving them against their bending.

BOOK IX.

Of the CULTURE *of* TARES.

CHAPTER I.

THE SORT OF TARE.

THETCHES are of several sorts, as the great horse-thetch, and the smaller sort, which by some are called the winter-thetch; these are very profitable if sown in a right soil, and at a right time: the horse-thetch was sown by a great farmer at *Penly*, in *February*, and only harrowed in upon his fallow ground on one ploughing, for his horses, cows, or sheep to feed on the following summer, either in the rack or field, and the remainder to plough in, a fortnight or month before they sow their wheat; on this they harrow their wheat, and dress it with short dung, or fold on the top. The smallest West-Country winter-thetch is sown in *September*, for food for the sheep, &c. in the winter and spring: they first sow them broad-cast all over a

piece

piece of chalk, gravel, sand, or loam (for wet ground is not proper for them) and then plough them in under-thorough, where they will make their way through the ground, as being a most hardy grain, according to the comparison of an old saying—*A thetch will go through, the bottom of an old shoe.*—And therefore many sow them among pease, because if they miss, the thetch generally hits: their dry haulm is but coarse fodder for horse or cow, but the corn is good for pigeons, and to give store-hogs: three bushels sow an acre, and often return twenty.

LENTILS, are the smallest sort of pulse that are sown, and lesser than the thetch, as that is the pea, and the pea the bean: they are sown on the poor chalks, sands, and gravels, where neither the thetch nor pea will thrive; there this will flourish and produce great quantities on its upright stalks, that grow about a foot and a half high; are sowed alone, or with oats, and make the best of provender, given in the rack as it comes out of the field, which the cows and bullocks will greedily eat, and fatten very fast under its keeping; swine are also great lovers of the lentils, and will pick up what falls on the ground, and be much forwarded in their flesh; pigeons are great lovers of them, as being very natural to their bodies. In many places they sow these for their horses and cows instead of hay,

Chap. I. THE SORT OF TARE.

and commonly put chaff into their mangers, for the lentils to fall amongst; as the horses and cows pull the haulm out of the racks, that supplies the want of hay and oats: one bushel sows an acre; and generally returns fifteen. They are often sold for three shillings a bushel.

March is a very proper time to sow lentils in, either alone or among corn; they very much affect to grow in dry soils, and therefore are commonly sown among oats; half a bushel of lentils, in this mixture, is sufficient to sow one acre of ground: if sowed alone, a bushel and a half is but sufficient to sow over one acre; and, if a favourable summer follows, there may happen to be fifteen bushels of lentils got off the same. If they are sown with oats, they are easily parted from them in the barn; because, on throwing the oats and lentils together, the lentils, being the heavier corn, will fly further than the oats, and so may be separated; they are easily mown, as they hang and twine together; and what is very profitable to this grain, they will grow well in poor dry soils. It was these lentils in some places that stood the poor's friend in the hard winter and spring-time of 1739-40, when many had lentils ground down into flour, and mixed with oatmeal for making bread, when wheat was sold at seven shillings a bushel in
Hempstead

Hempstead great market. Near *Brackley*, about twenty years ago, some farmers used to sow lentils in a whitish soil among barley, which, being ground together, made a bread in dear times of wheat for the poorer sort of people. Some are of opinion they beggar land when sown alone, because they seldom afford a full cover to it.

Winter tares, in *Hertfordshire*, are called *Thetches*; in *Middlesex*, *Tares*; in some other parts *Vetches*. One author gives vetches four several names or distinctions, as the *Gore-Vetch*, *Pebble-Vetch*, *Winter-Vetch*, *Rathripe-Vetch*; of all which I shall only here take notice of the winter-thetch or vetch. This is not the largest but the hardiest thetch of all others, therefore most proper for what I am going to recommend it; and that is, for sowing it about *Michaelmas*-time as a most valuable piece of husbandry, because, by so doing, these thetches will come in for feeding horses, cows, and sheep in the spring-season, after turneps are gone; and this is one great benefit belonging to inclosures, for here we can sow and time a crop of grain, thetches, or grass at our pleasure, when the common open field denies us.

A farmer who occupied about two hundred acres of land in *Studham* parish, *Hertfordshire*, being very desirous to enjoy the benefit of this sort

Chap. I. THE SORT OF TARE.

sort of thetches, was as the charge of sending for the seed as far as *Wickham*, in *Buckinghamshire*; and, having got the right winter large sort, he plowed up a wheat-stubble inclosed field, that lay in two bout-lands, into broadlands, and harrowed in two bushels of them on an acre about *Michaelmas*-time. These grew into a most fine crop by the help of a mild winter, so that he baited his store-weather-sheep on them for a considerable time, from the month of *May* forward, thus: In the morning his shepherd drove them from the fold to the common, where they remained till about two a clock in the afternoon; then he brought them into the thetch-field, for filling the bellies against folding-time, and they would carry such a quantity of this pleasant wholesome juicy food away, as caused them to dung and stale prodigiously; so that the land was almost double dressed, in comparison of that meat got only from off commons.

The next year this farmer sowed the same sort of thetches again upon one of his stubble-grounds, but missed of that success he the year before enjoyed; for it happened that a most severe frosty winter followed, that perished his whole crop, for at the spring they looked reddish, as if they had been singed, which made this farmer plow up the same ground and sow it with peafe; which fo difcouraged him, that he never would venture to fow any more winter-thetches, but betook himfelf to the fowing of clover and trefoil for his fheep, under pretence of their being a more fure crop However, this is not the cafe of many others, for now thefe winter thetches get more and more into ufe, for their forward and great fervice, and that for almoſt all forts of farmers cattle. And, though I have wrote that this farmer fowed them on only plowed up ftubble, yet many make a fallow on purpoſe for them, that their ground may be got fine and hollow enough, to caufe a furer and fwifter growth of them.

The winter-thetch is very valuable to fow at *Michaelmas*, not only for horfes, cows, and fheep, but alfo for obtaining a very forward crop of them for feed; for, by fowing thetches fo early, they will be ready to mow early, and be got into the barn in the drieft and hotteft feafon, and thereby give the farmer an opportunity to fow the fame ground with turneps, or *French* wheat, cole-feed, or indeed common wheat; for hardly any vegetable prepares the ground better for the reception of feed than the thetch, and this it will do to admiration, infomuch that many farmers think a full crop of thetches will fo kill weeds, and enrich and hollow the earth

by

by their great cover, as to equal several plowings.

The gore-thetch is preferred by many for a crop to feed horses with in particular, while they are in their green condition; for these sort run into a large and very long stalk and kid, beyond all others, and are therefore fitter for feeding great cattle than the smaller sort of thetch; but, as these are more tender than the winter-thetch, they are seldom sown till *February, March*, and *April*, and then they rarely miss of a plentiful return; if the ground was tolerable fine and in good heart; for a thetch is a very hardy vegetable, and of great use to a farmer, because they will not only supply his horses, cows, sheep, and hogs, with meat in the field, but also in the stable, cow-house, and stye, if they are daily mown and given them. It is the practice of a great farmer, near me, to mow his thetches as soon as they are grown into a good head, or in blossom, and till they are in kid, but not when too old, and gives them to his hogs; for some sort of swine will eat them greedily, if they are cut and given them every day fresh, and so thrive as to become pork, if they have no other food. I sowed about two acres of a large field with thetches, for my horses, and, though the rest of the field was plowed several times and dunged the same as this piece was, yet I had better wheat where the-thetches grew than in any other part of the same field, though the same was not plowed so often as the rest was, as it was proved in the year 1741. It was this sort of thetches also that maintained my horses alone, under the work of cart and plow in 1742, good part of the summer, and are of such a fattening nature, that a horse, in three weeks or a month's time, will get fat with them. This piece of husbandry is also performed in many open fields, as it is this year 1742, among *Lent*-grain, in *Edgborough* common field, in *Bucks*, where, a farmer of my acquaintance having but one piece of ground in one part of the field, and another at a distance in the same, he runs hurdles along the outside of the piece, and then baited his store-sheep on them for folding good part of the summer.

Thirdly, In the *Chiltern* country we sow them in two bout-lands, by plowing them in; or, *Fourthly,* On broad-lands, and harrow them in. This is giving a practical account from the result of experience, which would have been impossible for me to have done, had I not travelled for a great deal of my knowledge; so that, however deficient I was formerly in my writings, I hope I am now able to give my readers that satis-faction.

faction as may tend to their profit in particular, and to my country's interest in general.

The large thetch, or vetch, as it runs into a large, long stalk and kid, covers more ground than the small thetch; and, the more ground it covers, the greater benefit it receives. On this account, some farmers are of opinion, that a large thetch does the earth as much service again, as a very small thetch; because its weight, close lying, and large stalks very much enrich the ground, keeping the spirit or vapour of the earth in it, defending it from the damage of drought, and exhalations of the sun, lodging quantities of dews, preventing the growth of weeds, and killing others that had before their roots remaining in the ground; which the smallest thetch is not so capable of doing, because its weaker roots, lesser and shorter stalk, and kid yield not that cover as the larger thetch. It is this larger thetch, that is called the winter-thetch; which therefore admits of its being sown about *Michaelmas*, in order to acquire an early strong root and head, to enable it to withstand the severity of the winter weather, and get into such a growth in *March* or *April*, as qualifies it to become a spring-green grass-food for sheep, or sheep and lambs, that, in this scarce time of green meat, may then enjoy plenty of it, and feed the thetches twice over; which will greatly

greatly improve the land the thetches grow on, by the dung-ftale, and oily wools of thefe beft of quadrupedes.

Wild thetch is alfo an excellent natural grafs, whofe long ftalks in good land, and in fertile feafons, are loaded with many little kids, that contain numbers of fmall hearty feeds.

CHAP. II.
TARES ON A SAINFOINE LAY.

A Farmer that was owner of a chalky, loamy field, that had fome time been under fainfoine grafs, till it was in a manner worn out, took this courfe, to make the beft of his ground, viz. he ploughed up his fainfoine field with the ftrong double-fhare plough, by fix horfes that went three double, by which he ploughed up very narrow thoroughs, or furrows, in this fhort, foftifh, chalky foil, that proved a great advantage to the fowing of the fame with corn; for, as we fay, the more thoroughs, or furrows, the more corn; and this double plough makes more thoroughs in a field, than any of the fingle ploughs can, and lays them much evener, and better for the reception of the feed, and nourifhing it afterwards; and when all the field was thus ploughed into broad-lands, he harrowed three bufhels of thetches on each acre, and they throve fo faft,

by

by the freshness of the new broken up grass-lay, that he began to mow the green thetch-crop in *April*, for feeding his plough and waggon-horses with it, out of a rack in the stable, and so continued mowing, and feeding, the thetches, till some time in *May*; when they were thus all mown off, he ploughed the same ground first into broad-lands, with the two wheel single fallow plough, and when the land had lain in this posture a little time, he harrowed it all thoroughly fine, and then directly hacked it across the last way of ploughing, shattered the earth, and reduced it into a powdered, porous condition, fit for sowing it with any suitable grain: and when the field was thus got ready, he went to *Watford* market, and bought a parcel of the white *Essex* roading-pease, at five shillings a bushel, which being a small sort, a few of them went a great way, in sowing this field; and the further, for sowing them in drills, at three feet and a half distance; and at the same time he sowed turnip-seed in the interspaces between the drills of pease, which furnished the whole ground with seed, and then made one harrowing once in a place serve both. Afterwards, when the the turnips and pease were got big enough to hough, he made one trouble serve for all, and houghed both pease and turnips before he left off.

CHAP. III.

OF THE TIME OF SOWING.

IN *Hertfordshire*, we call them thetches, that may be sown for an early crop about *Michaelmas*, and then they are called winter thetches, because of their early sowing, and feeding sheep and horses in the spring-time, for which use, they are exceeding profitable, as coming at a season, when all other green meat is very scarce: but the thetches sown at that time of year are a hazardous undertaking, for I knew one of our top farmers sow them about *Michaelmas*, and he succeeded to his wish, by having a very early crop of them. But, on a second attempt, the following year, at the like time of sowing them, he lost the whole crop by it; for, though they made a fine appearance all the next winter, which was a mild one, yet, at *Candlemas*, they were killed by the severity of frosty weather, so that he was discouraged from ever sowing them afterwards at that time of year, and betook himself to a later and safer way, which was, to sow them in *January*.

First, if you sow thetches in *October*, *November*, *December*, or safer, in *January*, in a *Chiltern* country, sow three bushels on an acre; for, when thetches are sown in any of these winter months,

Chap. III. OF SOWING.

months, they should not be sown in less quantity, and then they will be but equal to two bushels and a half, sown in *March* or *April*: for, in these two spring months, the thetch is, in a manner, out of the power of violent frosts; and the more, as it is, in its own nature, a hardy, hot vegetable; yet not so hardy, as to be frost-proof, when it is a very severe and long one, which seldom happens at that time of year. Now there is a winter large thetch, and a small lenten thetch. The first sort are sold in common at *High-Wickham* market, in *Bucks*, and at other places. The last are much sown about *Warminster*, in *Wilts*, where they call them lenten thetches, being a small, round sort, sown in *March* or *April*, and therefore called by that name, and generally return a very great crop.

If any person have a mind to sow the large winter thetch, they should do it in a gravel, sand, chalk, or other dry, warm soil, and where such land is screened from the fury of north and east winds, by a low situation, or by tall, thick hedges, or good; for, if a farmer should be so indiscreet to sow the winter thetch in a clay, or other wet, cold soil; or, in a gravel chalk, or sand, that lies full bleak, and exposed to the power of a north, or east wind, he has reason to fear the ill consequence of losing his crop of thetches by it.

Secondly,

Secondly, Thetches may be sown in divers manners and forms. In vale, clayey, or wettish grounds, they may be sown, and then ploughed in, as they do their wheat-seed, on their ridge, rood, half-acre, or acre-lands: or all the thetch-seed may be harrowed in on the rough ground, as soon as the plough leaves it. And here the thetches, sown in this month, will be of the greatest service for staking the cart and plough-horses, in *April* or *May*, in common fields, as is very usually done by some of the considerate husbandmen, who can foresee the want of this delicate horse-food, before they feel it.

The month of *February* gives the farmer a most valuable opportunity for sowing thetches, in order to obtain an early and large crop of them, because it is most likely, in this month, for the seed to have the benefit of rains, which admirably agrees with the thetch, for wet weather will not kill a thetch; but much dry weather is an enemy to its posterity in its infant growth, before it is got head enough to shade its roots.

CHAP. IV.
OF THE QUANTITY OF SEED.

WHEN tares are sown after wheat or barley on ridges, two bushels and an half of the smaller or larger sort will do; but if on broad land there should be three bushels of the large sort sown, and two and a half of the small, and harrowed as soon as sown.

CHAP. V.
OF THE APPLICATION OF THE CROP.

IN *May*, those thetches, that appear speckled or begin to blossom, may be daily mowed, and given in racks, as we do green clover, &c. to horses in the stable, for soiling them; and, as they are of a very healthful nature, they cool their heels, keep their bodies open in hot weather, and suddenly fat them. Thus such a crop has lasted three weeks cutting, before they are too ripe, and the sap or goodness out of the stalk.

Thetches, fed or mowed green, are a great great improvement to the earth; for, *First*, they employ the fallow ground. *Secondly*, the great cover of the spreading thetches keeps in the spirit of the land, and thereby very much enriches it. *Thirdly*, The horses dress it with their stale and dung. *Fourthly*, They kill weeds, and so hollow the ground, that, on one, two, or three ploughings, wheat may be sown, as usual, in *October* following.

Thetches are also good to mow and feed cows in racks; for, as they are a green food, they will breed abundance of milk, while the beasts are freed under cover from the torment of flies, and the scorching heats of the sun; and, to enjoy this great benefit the longest time, thetches should be sown in *January*, *February*, and *March*, or longer, that their alternate growths may seasonably furnish the farmer, both in vale and *Chiltern* lands, with a sufficient quantity of green meat, when no other sort is to be had; and this not only for horses and cows, but also for sheep, and especially those ewes that suckle house-lambs. Then, and in this manner, green thetches may be mown all *June*, *July*, and *August*, and be either given under cover, or fed in the field, to very great advantage.

But to further explain this, I am to observe, that, when thetches are big enough to feed storesheep with in the field, the farmer begins to set up his fold in the same on purpose to fold his sheep that feed on these thetches. This he fails

Chap. V.   OF THE CROP.

not to do every fair night after the sheep have been fed on the common, or other field, about half the same day, and the rest of it fed and baited in this thetch-field, till they get their bellies full, for enabling them to dung and stale in a plentiful manner: and, that these sheep do no more harm in the field of thetches than what cannot be helped, the farmer acts the good husband, and runs a row of hurdles a-cross it, to confine them to their due bounds; that, when one part of the thetches are fed enough down, he moves them farther to give them a fresh bite, and so from time to time till the whole field is fed and folded over.

Thus a farmer enjoys a plentiful dressing in the cheapest manner possible, even to a double profit, one by the feed, and the other by the dungs and urine of these excellent creatures sheep; creatures whose excrements agree with almost all sorts of land, that thus may be improved to a very high perfection both early and late: that is, this profitable piece of good husbandry may be carried on from the month of *May*, to near *Michaelmas*, by feeding down alternate sowings of this thetch-seed; which gives a farmer early and late opportunities of enjoying the profit of their several gradual green crops for his different sorts of cattle. If he is to feed his horses with them in the stable, how valuable is a

field

field of them that is situated near home? For then the servant can mow them every day, or every other day, and bring them home fresh for feeding his team of horses with them, that will thus enable them, with a little corn to do a great deal of hard work; and, at the same time, keep them in pure health; for the green thetch, thus given, will keep their bodies open, preserve their wind, hinder the swelling of their legs, and the cracking of their heels; prevent farcies, mange, and surfeits; and, in short, nourish these serviceable creatures to that degree, that with good management they may be kept in good flesh, and in good heart, while they labour early and late. So may cows be fed with this excellent green food in the same manner horses are, even till the thetch gets into its kidding or podding growth, and be given to them in the cow-house, where they may feed on this luscious green meat, during the hot summer season, and at the same time be delivered from the teasings of the troublesome, painful, biting, fly, and the scorching heat of the sun, which, when these creatures feel in excess, while they are confined in an open field, it fatigues them to that degree, as to lessen their quantities of milk, that, in a considerable number of cows, must amount to a great loss; and although it may be objected, that this is a troublesome and chargeable way, thus to mow green thetches,

Chap. V.   OF THE CROP.

thetches, and give them to horses and cows under cover, it may be well answered, that, by a farmer's so doing, he reaps several advantages, which he would not enjoy, if the horses and cows were fed in the open field. For, first, by their being kept up in the stable and cow-house, they are prevented trampling down and spoiling almost as much green meat as they eat, which both these quadrupede sorts generally do, by their running about, endeavouring to free themselves of their fly enemies. Secondly, the farmer, by this means, enjoys much more profit by their milk; for the succulent large stalks of green thetches are so juicy, when fresh cut, that they produce abundance of milk, while cows feed on them in a cow-house. Thirdly, they are hereby delivered from the torment of aking feet, which is a misfortune inseparable from those cows who are daily drove on hard ground to distant fields; for, by such a drift, the cattle's feet are made sore, even to a lameness, by reason they sometimes are cut by the sharpness of stones, or if not cut, the gravel that they take in and lodge, will not fail to give them some pain; and then the consequence is, that such a cow falls off her milk, and, withal, perhaps, to a great degree of loss; a loss that I have too much experienced myself, while I was necessitated to drive my cows to some
distance,

distance, before they could arrive at the field of grass or thetches; for cows may be fed on the green thetch, while it is growing in the field, as well as horses; but, for these reasons, it is much better to feed them under cover, if the near situation of a field will admit of it; and, if it does not, I am sure that a small drift, whether it be on hard or soft ground, will do them some harm. Fourthly, cows are, by this means of feeding them on mown green thetches in a cow-house, free of the danger of hoving, which is a danger that all cows are liable to, that feed on them in the field; a danger that exposes a farmer to the loss of all his cows in one hour's time or less, if they feed on green thetches in a wet day, and in a high growth of them, with a very hungry appetite; because their sappy stalks are then loaded with liquor and wind, which, if taken into their bodies in too great a quantity, may probably hove and burst them. Lastly, when either horses or cows are fed daily with mown green thetches under cover, they will make abundance of dung, which though it is a soft sort, and not so good as that produced by the feed of hay and corn, yet it may be made to do the farmer great service, if he saves it in a right manner, and applies it as well; that is, if he preserves it from the wash of rains, mixes it with harder dungs, and lays it on gravelly, sandy, chalky, or other

dry,

dry, hot foils, where fuch cool, greafy dung will do moft fervice. And how precious a commodity all dungs are to a *Chiltern* farmer efpecially, who rents feveral fields of hungry foils at a great diftance from towns, which incapacitates him to receive any benefit from buying dung, fo far off, when he thus enjoys it at home in the cheapeft manner poffible; which leads me to obferve further, that there is no piece of hufbandry in all the virgilian or old way of farming, that exceeds this of fowing thetches, and feeding their green crops off with cattle kept in the field, or houfe; for, in either way of feeding them, they do the ground great fervice, becaufe thefe, like pea crops, prevent the breed of weeds, and kill others that are old poffeffors of the field, by their great and clofe cover.

They likewife, at the fame time, hollow the ground to that degree, that one ploughing of it afterwards for fowing the fame with turnips or rapes, or wheat, will do where two would not, if a crop of thetches had not preceeded their fowing; and when a crop of green thetches are eaten or mown off the land by *May*, or the beginning of *June*, fuch land may, by only one or two ploughings at moft, be brought into a fine tilth, fine enough for receiving turnip-feed of the forward fort, or a late fort; if the forward fort are fown, as the feed of the *Dutch* turnip, they may be drawn or fed off time enough to sow the same field with rape-seed; and, after these are done, a wheat-crop, or a barley-crop, may be set on the same; and all this performed without the help of carrying any dung or manure to the field, provided such thetches, turnips, and rapes are fed off with sheep; for by this means the ground will be full rich enough to carry forward any of these after-crops to great perfection, because the weeds will be crippled, and the land plentifully stored, and furnished with the nitrous qualities of the sheeps dung and urine. But the profit of feeding green thetches is more than ordinary known to the farmer that suckles house-lambs. These enjoy their benefit by their ewes that feed on them in the field, and, by this juicy food are capacitated for a long time to give abundance of milk, that nourishes more than their own lambs; for green thetches will perform all this, when natural grass cannot, because, when this is dried or burnt up by the violent heats of the sun, the green thetch grows fast, as being secured in its roots by the shade of its stalks from this misfortune: and, indeed, in this and some other respects is of greater value to a farmer, than either clover, trefoil, or saintoine grasses; for, if these artificial grasses are fed down too close by the sheep, their stalks will bleed, or spend their sap so freely, as to cause

their

Chap. V. OF THE CROP.

their after-shoots to grow up weak and late, if not quite kill them; a misfortune that the farmer is not in danger of from his feeding a crop of green thetches; because, when these are fed bare, he ploughs up the same ground, and then there is an end of the thetch-crop. This long discourse on the profit of sowing thetches I write from the field of practice, because I every year sow this seed myself; and what I have wrote of the same, is well known to be the words of truth by those farmers who do the like.

The green thetch, mown and given to horses in racks, is an excellent soiling for them, will fat them very expeditiously, and is one of the wholesomest of vegetables to this serviceable beast. A mare fed on them, a few days, will be ready to take horse. A cow fed on them, in field or house, will give abundance of milk, and fatten at the same time. Nothing exceeds the feed of the green thetch for sudden fattening of sheep, making ewes fatten lambs a great pace, and if store-sheep are fed on this green thetch, it will cause them to dung and stale very much in the fold, to the great improvement of the land; on which account, either fat or store-sheep may be kept in the thetch field, even till it is in blossom, and a little time afterwards; in short, I feed my plough-horses with these green thetches or vetches, and find they work under their feed with pleasure; but, after the first mowing or cutting of them, they do not rally, as we call it, *i. e.* they do not grow again to much profit. When thetches grow into a thick crop, they require good weather to dry them in; therefore, those who sow thetch-seed late, are in the wrong of it; for I have known such forced to mow them for a corn crop in *September* and *October*, before they could venture to cut them down for the rain, and then they proved good for little; for a crop of thick grown thetches is very hard to get dry, after being thoroughly wetted.

In *June*, the thetches sown in *March* or *April*, are likely to be in bloom; then, as soon, as the forwardest are so, begin to mow them for feeding horses in the stable, and they will not only produce a great deal of dung, but fat them very suddenly, and keep them in health. By thus letting your thetches grow to a large cover, they will kill weeds, and so hollow the ground, as to become an excellent preparation for sowing the same land with wheat, on one or more ploughings, with good manure. Others will let them grow to a length, and after rolling them flat, will plough them in as a dressing for sowing and harrowing in wheat on only one ploughing, as *French* wheat is done for the same purpose, and it will prove a great assistance, especially, to all sandy, gravelly,

Chap. V.   OF THE CROP.   421

velly, and other such soils; but this should be done, the latter end of *September*, that the green haulm may have time to lie and ferment, and rot in the ground, for two or three weeks, or more, before the wheat is harrowed in. Or, if the thetches are not too forward in their growth, it may be better done in *August*. Others that sowed thetches, for feeding their store-sheep on them in the field, will enjoy this vegetable in a very great perfection, because this feed will this way create a great deal of good dressing by the fold, and at the same time keep the sheep in rare heart and flesh, when grass in fields, and on commons, is scorched up.

Thus, by sowing thetches in several pieces of ground, in one or more fields at different times, they will be ready for this use in *May*, *June*, *July*, *August*, and *September*: an improvement that is much put in practice of late, both in vale and *Chiltern*, and like to be more and more, since it is not only exceeding serviceable to the ground the thetches grow on, but also to several sorts of cattle, in a time when no other green meat, perhaps, can be got.

In *June*, thetches are in many places just in bloom; or, if they have begun to pod or kid, may be fit to mow for making hay of them: to do which, cock them in little wads as we do clover-grass, and, after two or three turns, they
may

may be hayed enough to carry into the barn. The reason why some mow them while the leaf is on their stalks, and before they are ripe, is, because the leaves, and kids, being all green, become a rare food in frosty seasons for horses, cows, or sheep, who will eat even the very stalks up clean, to their quick encrease in flesh and courage, for such fodder will supply both oats and hay: whereas, if they were to stand till ripe, the leaves would fall off, and the cattle eat nothing but the kids, because the stalks then would taste bitterish, and be refused. This is an excellent piece of husbandry, though observed by few.

I call thetches green, because it is here meant, thetches growing before they are in blossom, or all the time they are in blossom; and, when in this condition, green thetches are made use of, for feeding horses in the field, or mowed daily, and given to them in racks in the stable, they are of excellent service to the farmer, because they fat horses suddenly that do not work; or, if they are fed on them under their ploughing and carting, they will prove hearty and healthful food, by causing them to be leisurely loose in their bodies. It is these green thetches that give many farmers the profitable opportunity in summer-time of feeding their store-sheep, by baiting them on the same part of every day, after they
have

Chap. V.   OF THE CROP.   423

have been feeding on the common all the morning; which double feeding gives the sheep a full belly-full, and then they dung accordingly in the fold.

If thetches are sown early in *February*, in a right soil that is in good heart, they will be fit to mow for feeding cattle in the rack in *May*, or for feeding them in the field with this fattening green food. If mown for the rack, one acre of them will go further than two eaten in the field, by reason of the great spoil the cattle make in the field, by treading and damaging the tender stalks with their dung and stale, and at the same time greatly enrich the earth by their full cover, as I have before observed; which adds another valuable advantage to the farmer, for, by the good qualities of such a full crop of thetches, the earth is brought into such a hollow fine condition. It is usual with great *Chiltern* farmers, especially who occupy large inclosed fields, to sow thetches at different times, that one crop of thetches may succeed another to eat and fold upon; and when they have been thus eaten by the sheep, the same land may be ploughed up, and so again it may be ploughed once or twice after, in order to get a tilth for wheat. Others will plough thetches in, when almost knee-high, to lie and rot in and dress the ground for a wheat crop, instead of dunging it, and they
E e 4   will

will smoak and stink while they are putrefying, to the great enriching of the ground, for producing a large wheat crop. This is one of the best sorts of husbandry; for, by this management, a small inclosed farm may be made to feed a great many sheep.

Besides sowing thetches to be fed green by my horses and cows, I likewise enjoy their benefit by letting my folding sheep eat them in the field, thus; I hurdle a small parcel out, and, after my sheep have fed on the common till about two or three o'clock in the afternoon, I have them brought in to the field, and bait them on these thetches, which fills their bellies, keeps them in health, and half fats them; therefore I fold them in the same field every night, for dressing the fallow ground, and thus prepare it for the reception of wheat seed at *Michaelmas* following, to the saving a great expence, that I must be otherways at, in buying of manure to supply the same.

* * *

Threshing Tares, Oats, Pease, Beans, &c.

The thrashing of oats in *Middlesex*, *Herifordshire*, and in vale countries is performed all one way, but their cleaning is not; for, in the vale, they differ from the two first countries, by wind fanning their oats after this manner.

Chap. V.   OF THE CROP.   425

The Vale Way of cleaning Oats.

After they are thrashed, they rake and cavin them, to get out the short straws, and other grofs foulness; which when they have done, they are ready for the wind-fan; and accordingly the wind-fan is brought, and two men and a boy employed to clean them; some fan the chaff like kernels, and trumpery out of doors into the yard; others, that are better husbands, fan all their oats so as to save their chaff in the barn with the lightest kernels. The tail, or light part of their oats, they sometimes fan twice.

The Way of cleaning Oats, in Middlesex and Hertfordshire.

Their way of cleaning oats is much the same in one county as the other. They proceed thus: As soon as the thrashing part is finished, they rake off the short straws from the heap, and pass the rest through the cavin large holed sieve; when this is done, they fan them with the knee-fan; then pass all through a ridder-sieve, which lets out the best oats, and causes the trumpery to rise up to the top, which is taken out by the hand of the sifter, and put into a place by themselves. Next, in case the oats are thus almost cleaned, they only flack them with the knee-fan again, to take out all the remaining feeds of weeds, very light kernels, and other trumpery,

pery, and so bushel them up into a sack; but, if they were not so clean as to do with what I have wrote, they pass them a second time through the riddler, or riddle-sieve, and then they are commonly done in perfection.

The Vale Way of cleaning Beans, Pease, and Thetches.

After these are thrashed, the heap is raked, and both corn and chaff passed through the cavin-sieve: when this is done, any of these grains are ready to be cleaned by the wind-fan, which will blow away the chaff and trumpery sufficiently for sacking them; for they seldom fan these twice.

The Middlesex and Hertfordshire Ways of cleaning Beans, Pease, and Thetches.

Pease and thetches are commonly cleaned thus: As soon as they are thrashed out, the straw and trumpery are raked off the heap; then they are throwed by the casting shovel, to take out and separate the chaff from the pease and thetch: next, they are passed through the wheat ridder sieve, for clearing them of burs, small kids, and such like offal; which, by the riddling, or round sifting, causes them and the rubbish pease or thetches, to rise on the top, and so are taken out by the sifter's hand. They never throw these twice, because they are of the

heaviest

heaviest sort of corn, and therefore the easier separated from their chaff at once throwing. Horse-beans are cleaned in the same manner, as pease and thetches in these two counties, and several others. Two men are seldom seen to clean corn at one and the same time in these counties, unless the floor is a very large one indeed. One man can fan and clean five quarters of oats in a day; for these are of too light a body to clean by throwing. And one man can clean fifty bushels of pease in a day, by once throwing and sifting them. I should have said, that the thrashing of beans, pease, and thetches is done all one way in vales, *Middlesex* and *Hertfordshire*; except that, in *Middlesex*, they for the most part bind up the horse-beans in sheaves, as soon as they are cut or reaped, and then they are lain on the thrashing-floor as we do wheat-sheaves ready for the flail;-and it is by this means that such horse-beans may be thrashed easier, and much sooner than in the vale, or *Hertfordshire* way, which is to spread them on the barn-floor promiscuously as they are mown and brought out of the field.

How some clean their Corn in Oxfordshire.

Their general way of cleaning corn, here, is by the wind-fan; but when there is an urgent necessity for cleaning a grist of wheat to be sent to the mill, a couple of women will sometimes take a sheet in their hands, which, by flacking to and fro, will clear the chaff from the corn; but this is a makeshift; for this sort of cleaning will not do for market. About twenty or thirty years ago, women in some parts of this country, as well as in many more, would exert their strength and skill in husbandry affairs, more than they do now; for then they would reap corn, fill dung-carts, drive a team, and now and then, for making haste, would mount a horse and ride away with the same to the mill or otherwise.

The Kentish Way of Thrashing Corn.

Here they are so sagacious, as to bind up not only their wheat in sheaves, but also bundle up all other corn; by which means the ears are laid in two rows, one against the other, and thus give the thrasher an opportunity of beating out more corn in one day, than when it lies promiscuously or in a confused manner on the floor. I should enlarge here on their wise method of first tying up their corn in bundles in the field before they house it; but, as I have been pretty copious on this matter before, I shall write no more here of it.

BOOK X.

Of the CULTURE of BUCKWHEAT.

CHAPTER I.

The advantages of sowing French *Wheat.*

FIRST, it is a triangular seed, in shape like a beech mast, and about the bulk of a small pea: It yields a great quantity of white nutritious flower, with which some of the poorer sort of people make bread. Secondly, It is very serviceable for feeding and fattening swine and fowls; for which purpose, in *Suffolk* and *Norfolk* in particular, they sow a great deal of this grain, not only for dressing their hungry, sandy lands, but for raising seed to feed and fat their turkies and geese, which every year they send to *London* in vast numbers. It also feeds pheasants, partridges, pigeons, yard-fowls, &c. very expeditiously. But I think it was villainously applied when

when given to a horse for suddenly fattening him, to sell and deceive the buyer, as it was done at a certain fair in the North; for, when oats and hay were put before him, the horse refused both, to the wonder of the beholders, and so continued for some time; till, at last, a cunning fellow, suspecting the bite, advised that some bread might be given him, and he greedily eat it; by which behaviour they discovered the matter, and found that this horse had been fed with bread made of *French* wheat. Thirdly, milch-cows will feed on the growing stalks of *French* wheat in the field, and milk very well on this succulent plant, which, as such, receives much assistance from the air, as all the very juicy tribe of vegetables in particular do. Fourthly, It is likewise excellent food for ewes or wether sheep, in the forward part of winter, if sown in this month or next. Fifthly, If mown in time, it will serve as hay for winter fodder. Sixthly, It kills weeds, and very much hollows the ground by its great cover, retains the dews, and keeps the earth moist for nourishing its shallow roots. Seventhly, Near *Norwich*, they manage their *French* wheat this way: With one plough, they plough it in for a dressing to their sandy ground, and immediately that is followed by another plough. The first plough turns the *French* wheat in, and the other turns a furrow of mould

on

on that, by which it is all buried almoſt at one and the ſame time; then they harrow in their rye or common wheat directly to great advantage, *French* wheat, as I have ſaid, is alſo ſown with weld-ſeed, to protect it againſt droughts, and the damage of weeds. If it feeds ſwine, they give peaſe after it, or pollard.

CHAP. II.
THE CULTURE OF FRENCH WHEAT.

MY field was a gravelly loam, that had a crop of oats laſt on it, and, on the tenth of *December*, I ploughed it with the wheel fallow-plough into broad-lands, which I let lie till the beginning of *March*, when I harrowed it plain, and then directly hacked it a-croſs with the ſame plough. In this poſture it lay till the 7th of *May*, and then I ploughed it again, and ſowed rather above half a buſhel of ſeed on the rough ground broad-caſt, and harrowed it in once in a place: When this was done, I immediately ſowed my other half buſhel, and harrowed a-croſs twice in a place; by which means I haled it or covered it from the fowls, and ſecured it better from the ſcorching heats, than if it had been harrowed firſt, before any was ſown. If you ſow this ſeed, to plough its crop in, for a dreſſing to common wheat, then you ought to ſow it the beginning of *May*, that you may have time enough to plough it twice or thrice; which they do in some light grounds, in order to mix it well after it has first lain and rotted; but where it is to have only one ploughing in, then it may be sown the latter end of *May*. In some parts, in their sandy land, they plough it in with a wheel-plough shallow, and then immediately a foot-plough follows, and throws up a second mold or sand upon that, and then harrow in their common wheat, to give it a deeper bottom. Some, to have a crop of this *French* wheat, will make a fallow and dung for it, as for other wheat. If it is sown for seed, it may be done the latter end of *May*, or in all the month of *June*.

BOOK XI.

Of the CULTURE *of* TURNIPS.

CHAPTER I.

OF THE SOIL FOR TURNIPS.

FOR near thirty years I have fown turnip-feed in my clayey, loamy, gravelly, and chalky fields; and find that the clayey loams return the largeſt and rankeſt-taſted turnips, and the gravels and chalks the ſmalleſt and ſweeteſt; though it muſt be owned, that a ſandy loamy ſoil produces the beſt turnip of all other earths.

It is true, that there are different ſorts of ſoils in vales; but the general ſoil is a blackiſh clay, or ſtiff black-and-blueiſh loam, as it is throughout the greateſt part of the fine fertile vale of *Ayleſbury*; a ſoil too fertile, in moſt places of it, for the growth of peaſe or turnips. If peaſe are ſown here, they will run ſo much into ſtalk or haulm, that they will kid or pod the leſs for it; and if turnip-feed is ſown in ſuch rich earth, they will run into ſtalks and leaves to ſuch a degree, that they will have the leſs roots for it: But admit they did apple or bottle well, and grow here into a larger ſize of roots, yet they are neither fit to be fed on the ſame land, nor to be drawn to be fed elſewhere. If an attempt ſhould be made to feed ſheep, oxen, or cows, with turnips, on the ſpot of ground they grew on, and in the winter, the cattle would ſo ſink in and ſtolch it, that they would eat their meat in miſery, and grow rather leaner than fatter. Beſides, in ſuch an earth and ſeaſon of the year, the beaſt would daub and dirty the turnip in eating of it, and conſequently make it an unwholeſome food, ſo as to breed the red or white water, and perhaps a rot. But this is not all the miſchief that attends ſuch a vale turnip-crop; for if, to avoid the laſt evil, the turnips are drawn to be fed on meadow, or other dry ground, there would remain behind ſo many hollow places or holes in the land, as would give the rainy waters room to make a lodgment in them; and as, by this means, ſuch holes become receptacles for holding and retaining water, it would ſoak through moſt or all the upper part of the earth, and the whole ridge-land would be greatly damaged by it.

CHAP. II.

OF PLOUGHING FOR TURNIPS.

THIS is not written in behalf of the vale-farmers, who rent such low, wet, stiff lands, as forces them to lay it up in a ridge posture, for avoiding the damage of inundations of waters; because it is not to the interest of those to sow turnip-seed, for reasons I have heretofore shewn, although the same is so much in favour of the *Chiltern* farmer, that, next to wheat-crops, I believe I may say, turnip-crops are the next chiefest profit, because they not only employ the land the fallow year, and lay it under a profitable crop, while it would otherwise lie idle; but, by the turnips being fed with sheep, the farmer obtains a dressing worth sometimes twenty shillings an acre; for such dressing oftentimes so enriches the ground, that with a little more help (and sometimes it will do the feat without any) it will return the farmer an excellent crop of wheat, or barley. But, as a full crop of these is not to be expected, unless a due preparation is made for the turnip-crop, a right *Chiltern* husbandman takes timely care to fallow, or, to be plainer, plow up his bean or his pea, or his oat stubble in *November* for the first time, in order to let the same land lie to be improved by frosts and snows;

for, if thefe happen in any great degree, it will be much the better for it; the frofts will fweeten and fhorten it, and kill the weeds, which perhaps hereafter would greatly damage the corn-crops, that are to fucceed. The fnows likewife very much contribute to improve the land with their nitrous quality; after the ground has been once plowed for turnips, we commonly let it lie till this month, before we plow it a fecond time, which is called the firft ftirree for turnips, in order to plow it a third time in *May*, or *June*.

But to be more particular in this my account, of preparing ground for a turnip-crop, whether it be a gravelly loam, a chalky loam, an intire fhort loam, or a ftiff loam; I fay, whether it be any of thefe, we in *Hertfordfhire* generally plow an oat, or pea, or a bean-ftubble up, the firft time in fingle bouts, or form the beft of all others, becaufe it lays the earth up in the higheft pofture that a plow can do it in, and thereby expofes it to the power of the air, and confequently to the frofts in the moft exalted manner, for the deftruction of the feeds of weeds, and fhortening and fweetening the ground; which it feldom fails to do, becaufe it has a whole winter feafon to do it in. Then, as I faid, in *April*, a judicious farmer will plow it a fecond time, by bouting it again off the laft bouts; and then, I believe I may fay, the whole furface of the land has been removed, or ftirred to perfection; yet that it may be made intirely fine and fweet, in the next month of *May*, this fame land is to be back-bouted, and thereby prepared for the work of the harrows, for thefe have not been made ufe of all along. Now, about the latter end of *May*, the harrows are to be employed in harrowing the earth plain, in order for the farmer to lay on it his rotten dung; which when fpread all over the fame, in *June*, he will plow it into the ground, and then the whole furface of the inclofed field is ready to receive the turnip-feed, that is to be fown and harrowed in.

Thus I have given one particular account of the procefs of preparing an oat, a pea, or a bean-ftubble, for a crop of turnips. But there are feveral other ways of doing it, according to the nature of the earth, and a farmer's fancy, which, if I was to write here, would take me up more room than I can at this opportunity fpare; and therefore I fhall proceed to obferve, that where a farmer defigns to get fo early a crop of turnips, as to draw and fell in *July* or *Auguft*, or fooner, he ought to prepare his ground accordingly; which leads me to write on farther particulars relating to turnip-crops.

Where peafcods have been gathered, or other peafe are ripe and cut off, there you may, in sandy or other light hot ground, give it one plowing, and harrow in turnip-seed. This is frequently done in many places, especially where the *Carolina*, masters, hotspur, and other forward pease are sown, by the three-wheel drill-plow; and thus by sowing turnips the latter end of this month, particularly the *Dutch* sort, you may have a crop of wheat succeed.

Manures, or Dressings for Turnips.

It is to little purpose for any to sow this seed, on three accounts. 1*st*, If the ground is not before-hand plowed into a fine tilth. 2*dly*, If it is not well manured. And, 3*dly*, If the turnips are not well hoed. I shall here only touch upon the second article, and that is, 1*st*, If you dung the land for a crop of turnips, and it be of the long sort, it should have been plowed in at fallow-time, either before *April*, or in that month, or in *May* at farthest, that it might have time to rot and mix with the earth; but if it be short, rotten dung, then it may be spread a little before the last plowing, and plowed in, for harrowing the turnip-seed on it. 2*dly*, Or you may spread forty, or better sixty bushels of lime over one acre, and when it is slacked, you may plow very shallow, and harrow your turnip-seed on the same. 3*dly*, Or you may, as a top-dressing (as we call it) set

Chap. II. FOR TURNIPS. 439

your fold, and run it over the field as faft as poffible, elfe the fheep will damage the young turnips, unlefs they are parted from them in time, by hurdles. *4thly*, Or you may fow ten bufhels of peat-afhes over each acre as foon as the feed is fown, or fow them and harrow them in with the feed. *5thly*, Or fow forty bufhels of flacked lime over the acre, after the feed is fown. Or, *6thly*, Do the like with foot. The laft five dreffings will keep off the worm, fly, flug, grub, and caterpillar.

To prepare a Bean-ftubble for Turnips.

It was bouted overthwart the broad-lands, in *January*: The beginning of *May*, it was back-bouted. The beginning of *June*, it was tho-roughed down. At the latter end, it was harrowed plain, then dunged, plowed, and the turnip-feed harrowed in. This was performed in a wet, loamy, flat field of land in the *Chiltern* country.

Sowing Turnip-feed in Drills.

In this pofture, turnip-feed may be fown out of a hopper of the three-wheel or pulley drill-plow, in fhallow drills, which may be afterwards clofed by hoes, harrows, or by two fticks, fixed behind the broad boards, which leave the drill under a fmall ridge; and this by only altering or fhifting the round cog inftrument under the hopper, that is proper to drop out this small seed. For this purpose, the land must first be got into a very fine tilth, by frequent plowings with the common plow, and, if you can, dunged some months before with muck rotted very short beforehand. The drills may be made from two to six or eight feet distance: If left very wide, then wheat may afterwards be sown in drills between the turnips. Here turnips will thrive much better, than those sown in the random way, because, in the driest seasons, and in the driest soils, they will come up in less time, tho' less seed is sown this way on an acre of ground, than when sown broad-cast. Also, in this drilling posture, turnips are prevented from burning, or growing yellowish in their leaves a long time; are easier hoed, better secured against the slug, fly, worm, and caterpillar, and from that common destructive turnip disease, which in the sandy grounds of *Norfolk* frequently happens to great numbers of acres in a season, and is there called—*Anbury*—and which in a little time reduces one acre of fine large turnips, worth thirty shillings, to be worth no more than five. By these plows, turnip-seed may be drilled on level ground, or on ridge-lands, and, by being hoed by the break and hand-hoe in the intervals, wheat may be sown in *September* to great advantage; after which the turnips may be carried off, and sold

or

or eaten elsewhere, that the same ground may either lie fallow, till barley or pea-season, or cole-seed, or artificial grass, may be drilled in the inner spaces to great profit. Or, I have known this done, in the month of *April* or *May*: You may sow white, blue, or hotspur pease in drills, at six feet asunder, and at the same time harrow some *Dutch* turnip-seed, sown broad-cast, in the intervals, or it may be drilled in, as I said before, and afterwards hoed; by which management, the ground may be all cleared at a time, to drill in wheat, or sow it in the common way. Sands, gravels, and dry and moist loams are the most natural soils for turnips, and so are some sorts of rich moist chalks; but the dry, shallow, hurlucky sort is too dry for turnips to thrive in, for here they will lie so long before they get into a third or fourth leaf, that they will be in great danger of being devoured by the fly and slug, unless the weather be very rainy. But this drill way of sowing, with good manure beforehand in the ground, will make such earth hollow and rich, and cause the seed and turnips to grow by the help of only dews, and the shallow cover of loose ground, to great perfection, but the drills must not be made too deep, lest the seed be buried; and for security, after some earth is pulled down by the hand-hoe on the first sowing,

more

more feed may be sown on that, and covered. Thus you will have two comings up of turnips, and the surer still, if you sow old and new turnip-seed, which will prove another security against the fly and slug, for, if one is eaten, the next may stand. Or, besides what I have written before of sowing peafe, or wheat, broadcast between the rows of turnips: You may drill the turnips at six feet asunder, and afterwards sow wheat in a drill in the middle; and after the turnips are either eaten off by the inclosure of hurdles with sheep, or pulled up and given elsewhere to cattle, peafe, or barley, &c. may be drilled in their room; so that both crops may be got off almost together. However, this drilling practice will not answer in too stony grounds, nor even in some very dry loose sands, because the sands will tumble down so fast after the share, that the drill will be filled before the seed can fall in; but, where such earth will stand long enough for the seed to fall in, it will answer to good purpose. Remember, that the richer the land, the later you may venture to sow turnips, either in drills or broad-lands, for, in such a soil, latter sown turnips will overtake those sown some time before in poorer earth; and, whenever you sow wheat after turnips, there seldom fails a good crop. Do not be so much afraid of burying

your

your turnip-feed in drills, as to let it difcourage you in this drilling-work, for it has been found that turnip-feed, like fpinage-feed, will come up at three or four inches depth; but, if you doubt this, you may, as I faid before, fow another parcel of turnip-feed out of the drill-hopper, or hand, on the top of the firft, and with the hand-rake, or with a fingle light harrow, you may cover the fame with a fufficient quantity of earth.

Plowing and Sowing Turnips in Vale Grounds.

The aforefaid operations, by diverfity of plows are not to be performed here, becaufe in thefe fituations they never plow but one way; and that is done fometimes by ridging up, and fometimes by cafting the ground down with the foot-plow. Here the land generally lies in acre and half-acre of lands, but for the moft part in the latter: Where their foil is a black or bluifh, rich, clayey loam, they dare not fow turnips in it; for, if they fhould fow turnips in this rank, dirty, wet, ground, it would be very likely attended with ill confequences; as, their running into little roots, and great ftalks and leaves; and, if it fhould happen, that they take well, and be a full crop, yet then it will not anfwer, becaufe the fheep, in eating them off, will ftolch the ground, daub the turnips,

eat

eat their food in misery, and perhaps, get rotten into the bargain; and, if they are drawn to be fed on meadow, or more dry ground, then there will remain such hollow places, that, in this stiff, wet soil, will prove so many receptacles for the lodgment of waters, which in course will sour the earth for years after, as I have known done, and greatly damage the succeeding crops; besides which, there will not be time enough allowed, for giving the land any more than one ploughing for barley; and then, I believe I may say with assurance, that there will be such a bad raw tilth, as to cause a very poor return at harvest, as has been proved in several places in the vale of *Aylesbury*, and even upon the very edge of it, in the parish of *Eaton*, as I shall in another place, more largely treat on. Yet is this caution not without exception, for in some vales, as well as the vale of *Aylesbury*, there are many proper situations and soils fit for sowing and eating off turnips on the same; as sands, sandy loams, sandy clays, whitish ground, gravelly loams, and intire dry loams, that will very well admit of this practice.

In sowing turnip-seed on broad-lands, regard should be first had to the after use the ground is to be put to. If you intend to sow the same land with wheat, then sow the *Dutch* forward turnip

Chap. II. FOR TURNIPS. 445

turnip, to be eaten off in *September* or *October*. If you would have them stand part of the winter, sow the green tankards or green rounds; but if the whole winter, then sow the red tankands, or red rounds, for your sheep or bullocks. Again, in case your ground is a wettish soil, here sow the sugar-loaf, or one sort of the tankards, for, the upper part of this turnip growing pretty high out of the earth with a thick skin, the sheep may have easy access to their food, and the water less power to hurt it: And thus you may manage your turnip-crop, by sowing the proper seed in a due season, for early or late feeding, for pulling up, or eating them on the spot, for sowing of wheat, or rye, or barley or pease after them, &c. *July* being commonly the driest month in the year, and the fly of the greatest power, it proves a caution to many farmers, not to sow their turnip-seed so soon as the ground is harrowed plain, but to let it lie a day or two afterwards, and then sow and harrow in the seed, because the fly is apt to follow the dust. When land is well folded, or dunged and plowed till it is got into a fine tilth, and made ready for sowing, do not grudge seed, for there have been thousands of acres of turnips lost for want of a pound more than is usually sown on an acre. My quantity is never less than

three

three pounds on one acre, and sometimes four, if the tilth be rough; the more the better, that there may be enough for the farmer, the slug, and the fly. Now our method of sowing turnip-seed on broad-lands is this: First, he begins to sow half a broad-land, which is about four steps and a half broad, by stepping along, within a foot of the outside, and at every second step, with the right-hand, he throws a cast from between his two fingers and thumb: When he has got to the end of the land, he comes all along the contrary side, and sows here, as he did at first, by which two cross casts, the ground is sown twice in a place, and so on till the field is all sown; then he immediately harrows all the seed in once in a place, and leaves it till turnips are grown into their fourth seedling leaf, when they should be hoed.

Where land lies low and wet, and will not admit of turnips to grow on broad-lands, then the farmer must alter his mode of plowing and sowing. If it is to be done, in two-bout lands, or what we call four-thorough stitches, then, when the ground is got ready for sowing, harrow down the stitches or ridges long ways, almost level, and then sow and harrow in your turnip-seed; for here your turnip-seed is to be soon broad-cast all over the land, by crossing the casts, and sowing five

stitches

Chap. II. FOR TURNIPS.

ftitches in breadth, at a time, and fo on, till the whole field is finifhed; and, by this operation, there will remain a fmall loofe thorough between the ridges, enough to drain off the waters from the higher ground, unlefs a flood of rain happens; for, by harrowing down the earth from the ridges, there will lodge a great deal of loofe mould, which will receive and bury the wets, fo that the turnips will lie much drier, and in a hollower condition, than if fuch land lay in broad-lands. If the ground lies in three or four-bout lands, it is, becaufe the fituation of it is very low, and wetter than that of the two-bout lands, and yet not fo wet as the ridge vale lands.

CHAP. III.

OF THE SORT OF TURNIP.

IN the sandy lands of *Suffolk* and *Norfolk*, they sow the moft turnips of any two counties in *England*, for feeding their horfes, and for feeding and fatting their *Welch* and *Scotch* runts and fheep, for which purpofe, they sow the cream-coloured *Dutch* turnips, the yellow, the purple, and the green, and the red forts as follows: *viz.*

The cream-coloured *Dutch* turnip is of a flattifh fhape, and of a middle fize; and more fown by gardeners than farmers, becaufe of their quick growth for an early market; for this fort will be fit to pull in eight weeks time after hoeing, if the feed is harrowed in forward in the month of *May*, or in *June*, provided the foil is of the dry warm fort, is well dreffed, and got into a fine tilth. And, though it naturally is a fmall turnip, yet, by growing in a rich earth, they will become pretty large, and therefore are fown by fome farmers, as a common field turnip, to fat their fheep, as the firft crop to be eaten off time enough to plow and fow wheat.

The yellow turnip is of the carrotty kind, and the fweeteft of all others. Therefore the *Suffolk* and

Chap. III. OF THE SORT, &c. 449

and *Norfolk* farmers sow some acres of them in their inclosed and common fields, for their horses as well as for fatting their sheep and runts, because this turnip is so luscious and fattening, that they give them as manger meat to save corn; for, with these and hay, or with good straw and chaff, they will hold to work well.

The green sugar-loaf, or tankard turnip, is one of the sorts sown in *Hertfordshire*, and many other places; has the thickest rind or skin of all others, and for which it has the most occasion; because this turnip stands the highest in the air of any, and therefore the most exposed to frosts, which it will wonderfully resist, and maintain itself sound, unless the hard weather continues long and violent. It is a sweet turnip, and well beloved by sheep and cows, growing very large in rich ground, and this more especially, because the cattle can easily come by a full bite of it. But there are some inconveniences attending this turnip, and they are these: If tankard-turnips stand too long they are apt to grow corkey, or rot, and make cattle pine away instead of fattening them: This turnip is one of them that should be sown early, for, if you sow it later than ordinary, it will run only into leaves, and not bottle; and, whenever you sow it, it ought to be in a deep soil, that its slender root may meet with the

better

better encouragement to strike a good way down, grow the faster, and stand the stronger.

The green round turnip, is an excellent sort, and of late more and more sown in *Hertfordshire*, and elsewhere, because the true turnip of this kind grows large and sweet, and is not so subject to be spoiled by frosts, as the green and red tankards are, and yet stands enough above ground for sheep and oxen to come at the greatest part of it; will bear scooping well, and its shell easily dug up. This turnip, as the last, ought to follow the *Dutch* turnip, as the second or next successive crop to that, and therefore should be sown the latter end of *July*. This is a general turnip, because it will prosper in most soils, and is as good for the kitchen use, as it is for feeding cattle in the field: This is the most general turnip that we sow for our sheep in *Hertfordshire*, at this time, and therefore allowed to be the best sort.

The red round or purple turnip, is a hardy sort of turnip, of the hottest nature, and therefore will endure sowing the latest of all other field turnips, because it will best withstand the severity of frosts and chills of waters. This is accounted one of (if not) the largest turnip that grows. If it is sown in *July* or *August*, it will hold good and fit for feeding sheep to *Lady-day*, and in some kind seasons, till mid *April*. For, the sake of their
growing

growing in stiff, wettish loams the best of any, for their appling and thriving in winter, when the tankards will run only into leaf, and for their growing the latest of all others, this turnip ought to be preferred.

The red tankard turnip.——Some of this sort are very excellent turnips; some will grow almost out of the ground, and therefore are much exposed to frosts, and so easily pushed down by cattle, that great numbers are spoiled by them. But a true sort of these are those turnips that grow with almost as much of their body in the ground as out of it; such, as these contain a great deal of meat in them, and, when a farmer is possessed of such, let him make much of them, and not lose their seed; as many have done in our parts, whom I have heard say, could they have some of the same they formerly had, they would not grudge to give five shillings a pound for it. The red turnip delights in gravels and loams. But this, as well as the green tankard, in hard weather, is pecked by crows, rooks, and pigeons, and then the rain gets in, and rots them.

The rat-tail turnip, so called, because of its piky root, which runs down a considerable length into the earth, has small leaves and a green ring round a little white turnip, is one of the sweetest of turnips, but too small to be sown in fields for cattle; therefore is much valued by some gardeners for kitchen uses. It is so hardy, as to grow in winter, when others do not, and may be sown all the summer long.

CHAP. IV.

OF THE FLY.

BUT I must here observe, that though horse, ass, cow, and hog dungs are pernicious in breeding the fly, yet they have this good property belonging to them, that when any of them are truly short, rotten, and greasy, and thoroughly mix'd with the ground, it so lightens and hollows it, as to give an easy passage to the young turnip fibres, to strike down and get easily into the earth, whereby they acquire such a swift growth, as causes them to out-run the fly's rapine, and get into bitter leaves before they have time to demolish the crop, if the seed is sown thick enough; for in poor hard ground, it is often the farmer's misfortune to lose them, because here the fly has full leisure to feed on, and devour, the two seedling leaves at his pleasure, as is evidently proved by the successful crops of those who have, in time, dressed their field with a due quantity of rotten dung.

CHAP. V.

OF THE HANBURY.

THE hanbury overtakes turnips sometimes, after they have got large roots, but most of all in sandy soils. In *Suffolk* and *Norfolk*, the turnip disease is very common, in dry summers especially, and destroys great numbers of acres of them in a year, by worms growing in little bulbs or knobs on the turnips, and eating into their hearts.

CHAP. VI.

OF THE TIME OF SOWING.

IT is the established maxim, that a fortnight before *Midsummer*, and for a fortnight after, it is the best season for this purpose, for, if turnip-seed is sown sooner in the field, they will get so rank, cory, and stringy before winter, or spring, that the cattle cannot eat them.

CHAP. VII.

OF HOEING.

ON this absolutely depends the obtaining a good crop. It is true, that I have known some of the worse sort of farmers content themselves, to save charge, with only harrowing a thick crop, with two of our common four-beam harrows, which will thin the turnips, and tear up the best as well as the worst plants. Others, where the crop is thin, will not hoe at all; but these are your afternoon farmers, as we call them: For, to enjoy a pay-rent crop of turnips, they must be hand-hoed in due time, that is, when they have made their fourth leaf a little substantial, to prevent their burning, or too much heating one another by their too close standing, otherways, they will become stunted, and grow to little profit. Now to do this work effectually, in the first place I make a bargain with my hoers, to go over the crop twice, the last time about a week or fortnight after the first, to hoe up and destroy those young ones that may have escaped the first hoeing; which is more than ordinary necessary, if the weather happens to be wettish in the first hoeing-time, for then many turnips will up-set, and, though hoed before, will grow again. For both which hoeings, at *Gaddesden*, that is computed to be twenty-eight miles from *London*, we pay five or six shillings for each acre of ground, and seldom or never more. But, where many acres are sown, it often happens, that the hoers cannot do all, or even the greatest

part

Chap. VIII. OF HOEING. 455

part of them, time enough to save the rest from
spoiling. In this case, the triangular hoe-plough
will dexterously thin several acres in one day,
and excellently well prepare the young turnips
for hand-hoeing with the more ease and safety.
If the large sort of turnips are sowed forward
to feed cattle, they ought to be hoed at 18
inches a-part; if later, at twelve: Yet allowance ought to be made for the nature of the
turnip; a slender long turnip will grow well
forward in the year, at a foot distance, when
some of the large round sort will languish, if
sown so near.

CHAP. VIII.

OF THE APPLICATION OF THE CROP.

ONE acre of well-planted turnips will fat
eight middling-sized weather-sheep, or
feed six couples of ewes and lambs, with good
management; that is, if racks of hay, or pea-
straw, stand constantly in the field, for the
sheep to brouse on at pleasure; for this dry
meat secures them from the rot, from the red-
water, and from hoving; misfortunes incident

to all turnip-feeding sheep, and that often prove fatal to thousands of those that are deprived of such a salubrious antidote; and, for this purpose, some are so careful, as to have their meat-racks thinly thatched over; others have them covered with thin boards, to keep their meat dry, for then it does the sheep twice the service of wet meat. Others will be at the charge of close-made rodded hurdles, instead of open hurdles, for keeping sheep from being pinched by drift-rains, and piercing winds: which, with timely shiftings of the sheep into fresh parcels, will, in dry land, fat them with great expedition. When oxen, heifers, or dry cows, are feeding on turnips in the field, for fatting, great care should be taken that they are not choaked; and, if a piece of turnip sticks in their throat, the quickest way is to thrust a hand in, and shove it down. Others will keep a thick, stiff piece of rope constantly with them in the field, for the like purpose.

Some farmers, will draw the turnips, and feed and fat an ox, heifer, or cow in the house, by giving it chopped turnips in the trough, and hay in a rack, and then such ox, heifer, or cow, never will drink water, because the turnip, of itself, is such an aqueous root, as very well supplies it: But there must be a good parcel of turnips employed in fatting of one of these

horned

horned cattle, for fome will eat half an acre, others a whole acre, according to their bignefs, youth, and degree of fatnefs beftowed on them.

 In the year 1742, a farmer having a good crop of turnips, would keep his breeding-ewes among them, till they lambed, partly out of neceffity, and partly out of obftinacy, for he would ftill venture on, and run the rifque of the iffue; but the confequence proved fufficiently prejudicial, to teach him better management hereafter; for fome of his lambs were hawled away from the ewes bodies, by meer violence, and others forced to be cut away by piece-meal; which, befides the lofs of the lambs, caufed likewife the death of fome of the ewes. I therefore here offer this caution to all turnip and cole-farmers, that they take their breeding-ewes out of the turnip and colefield, about a month before their lambing-time, left the lambs and ewes be loft, by letting the dams feed too long on fuch nourifhing food; for the grofs fattening turnip-root, and quickfeed cole or rape-head, is very apt to caufe the lamb to grow too big in the belly of the ewe, for lambing without help; which cannot be done, without extreme hazard of both their lives; which hazard would probably be prevented in a great meafure, if he had taken the ewes out of the turnip-field, in due time, and put them in his meadow, and fed them

now

now and then with hay, &c. But this would not go down with him, becaufe he reckoned, that feeding the fheep in his meadow, and with hay, was the greater charge, although the charge, perhaps, might have been upon a par at the fame time with the feed of turnips; for the more hay the ewe eats, the more turnips there would have been for his wether fheep, cows, or hogs.

In *January* it fometimes happens, that great frofts, or deep fnows, deprive fheep and bullocks from coming at the feed of turnips in the field; and which likewife prevents their being drawn, or pulled up, for hoarding, or for a fale at market. Now, therefore, the farmer's bad, or good hufbandry, is brought to the teft; if he has made a turnip hoard in time, by placing a confiderable number of them in a barn, or other fheltered place, clofe enough to keep out frofts, and other weather, from damaging them, then it is in his power to bring fuch hoarded turnips out, or carrots, or parfnips, for feeding his fheep, cows, bullocks, or fwine, rabbits, turkies, geefe, or ducks, with them under cover, or in an open place, fo that thefe can have full room and power to eat them raw, or prepared; becaufe, as thefe roots are not frozen, and as cattle in fuch frofty feafons have the greateft appetite, they will eat much, and thrive apace. Accordingly, this good hufbandry of hoarding turnips, carrots, parfnips, and potatoes, is more

duly

duly obferved in *Suffolk*, *Norfolk*, and other fandy parts of the eaft, than any where elfe in *Great-Britain*.

CHAP. IX.
OF SAVING THE SEED.

WHEN this is defigned, due care muft be taken, that the turnips are no ways bitten by any cattle, for that may prove an impediment, though many turnips have grown after fuch damage, when not too much, and bore good feed. About a fortnight or three weeks before the feed is full ripe, one or more men or boys, fhould be employed to keep the fowls off, for the leaft fmall bird will do a great deal of hurt here: This is fo material a point, that I have known a careful farmer fave fifteen bufhels of feed off one acre, when his neighbours that were negleƈtful, got little more than his feed again, and this meerly for want of guarding it in time. One man, with a gun, will guard two or three acres well. Turnip-feed is ripe about the beginning of this month with us, as may be known by the light brown colour of its pods. Some reap it with fickles, as they do wheat; but the beft way of doing it is with fharp knives, that, by their keen edges, cut the ftalks eafy, without jarring the ripe pods, as the

rough

rough teeth of the sickle will be sure to do; besides which, a man by the knife can cut all the stalks off a single turnip clear of weeds, better than with the other; and thus with a knife a man may cut down a rood a day.

The cuttings or rips are to be laid in rows, and in this posture kept some days abroad, to hay and dry; and, as this seed is very apt to shed, some bring their blankets, or barn cloth, or sacks to thrash it on in the field. Others, again, will spread a barn cloth on the bottom, and about the sides of a cart or waggon, and bring all out of the field into the barn, for thrashing out the seed directly: two acres of haulm, with its pods, has filled two bay of barning, that sometimes will yield thirty bushels of seed, and more. An acre of this seed has been sold for above four pounds; one bushel of turnip-seed returns sometimes two gallons of oil, by expression, for the wool business, or for burning in lamps.

BOOK XII.

Of the CULTURE *of* RAPE.

THE value of this most serviceable vegetable is known to few farmers in this part, where I live; and therefore few, very few there are, that sow the seed of it, because we lie too far from *London* to enjoy the benefit and advantage of it in the manner that some do, who live nearer that metropolis, and carry on the suckling of house-lambs. I know of none within a pretty many miles of me, that sow this seed, besides a nobleman's bailiff and myself. The nobleman's bailiff sows it for his master's store-sheep, for feeding his weathers and ewes in *March* and *April*, and I mine; for then we bait them a few hours in a day in the rape-field, as the only green meat we have left, because it is the case of many to be without any other at this time of the year than this of rapes, by reason artificial and natural grasses, as well as turnips, are generally eaten off before these months. But the main design of sowing rapes is to feed ewes

that

that fuckle lambs, either for fatting thofe lambs that are brought up in the houfe, or the field, and for the fervice of the kitchen. It is thefe rapes that are of fo hardy a nature, as to withftand the violence of frofts, beyond all the garden ware of this tribe; when cabbages, favoys, brockely, and the like were killed by the vehemency of the frofts, thefe have ftood found, and fupplied their room, as feveral acres of them did in the hard frofts of 1739, and 1740; and, had it not been for thefe ferviceable plants, there had many lambs been loft; therefore, whether in frofty or open weather, in the fpring months, the rapes are of great value, as being not only a moft fucculent plant, that produces a great deal of milk, but they likewife fatten fheep and lambs very expeditioufly; fo that no farmer, that is mafter of a proper foil in a convenient fituation, ought to be without the enjoyment of this excellent food.

This feed agrees beft with ftiffifh or moift loams, but it will grow well in dry loams. The management lies in fowing them accordingly. If rape-feed is to be fown in ftiff, moift, or wettifh loams, the ground ought to be fowed betimes, even fooner than *Auguft*, for reducing its furly nature into fine loofe parts, by the month of *July*, when this feed ought to be fown in the fame, that the rape-roots may enjoy good part of of the summer's hot season for forcing them to take large hold of the earth, and get forward heads against the trying winter's frosty seasons; by this the plants will meet with the best security against hard weather, and yet, for all such a forward sowing of this seed, that rape-seed, sown in a dry loam in the month of *August*, is oftentimes as forward in its growth, as that sown in a wet soil in *July*. Therefore whether rape-seed is to be sown in *July* or *August*, in a stiff or drier loose soil, the ground ought to be early and duly prepared, for the harrowing this seed in; and, as rape-seed is commonly sown on the fallow ground, if it has not been ploughed up before *August*, do it now, either by broad-land ploughings or bouting it into single bouts, or in what we call four-thoroughed lands, or in three or four bout-lands, or in broad ridge lands.

These are all the best postures that land can lie in, that is of this stiff nature, for the first ploughing of it, in order to get it sweet and fine; for it cannot be too fine for this small seed; because, in case a person was to attempt the getting a full crop of rapes in a stiff, sour, clotty soil, if a great deal of the seed is not buried so as never to grow, the plants will be the longer enlarging their roots, and getting a flourishing head against winter; and then, perhaps, when they are

are to be fed, there are only half plants or rapes; and what a stunted crop of rapes, or any other vegetable, yields, a farmer knows by the loss he sustains by his bad husbandry; but this seed is not always sown in tilth grounds; for, where the land is of a shortish, dry nature, it is oftentimes sown on oat, or pea, or wheat, or barley, or bean stubble on one ploughing only, and harrowing in the seed after harvest; and, when the weather proves propitious, there are many good crops of rapes got this way.

If the land is an intire, a gravelly, or a sandy loam, if it is well ploughed, and well dunged, it may be made to become a good crop of cole or rapes; but when such ground is to be sown with this seed, special regard ought to be had to the proper time, lest time, labour, seed and other expence, be most of them lost. Now if a person has a mind to get a *Chiltern* inclosed field take with cole or rape, and to stand throughout the winter in a right order or growth, he must not sow this seed till the middle of *August*; for if he does, the young crop is in great danger of running into seedy heads before the winter; and then the stalks will get hard and dry, and, in short, be neither fit for cows nor sheep, nor hardly any thing else.

Hence it is that cole or rape-seed may be said to be sown at several times in the year, viz.

In tilth clayey grounds, as I said, the latter end of *June*, and in dry tilth loamy grounds in *July*, are the chief times for sowing cole-seed in fields; but about *Watford*, *Rickmansworth*, and other adjacent parts in *Hertfordshire*, the suckling farmers, or those who suckle house-lambs, take care to sow some of their land with the puffin *Hampshire* kid, or *Cobham*, hog-peas: for as these are a forward sort, they sow them to come off by an early opportunity, for ploughing the same land, and sowing with cole-seed in *August*, or in *September*, at farthest. Others sow the blue boiling-pea, and the *Essex* white roading boiling-pea, for the same purpose. Those indeed that sow ormuts, masters, hotspur, and the great union peas, and other sorts, selling them green in their pods, may certainly get them off much sooner than any of these; but as they commonly grow in sandy and gravelly earths, for getting them into an early order to set on crops of turnips, to draw to sell, or to feed their sheep with in winter, I have no more to say of them here, but proceed to observe, that though cole-seed is sown so late as in *August* or *September*, after a pea-crop, or a barley or a wheat-crop, on one ploughing up of the same, and harrowing in the seed; yet so it happens, that a good crop of cole, by these means, is often got, to the great advantage of the owner; I say, to his great advantage, because such later-sown cole-seed may produce a crop that will answer to as much profit as any early crop; for when the first crop of cole is eaten off, that was sown in *June* or *July*, this last will come into use in *April* and *May*, and serve to feed cows, sheep, lambs, deer, &c. in the scarce time of hay and grass; which too often then happens to be the farmer's want, by having expended his dry meat, before the grass gets head enough to become a sufficient food. And it is also on this account that some sow cole-seed in *February* and *March*; so that this herb is now become such a field-plant, that it may be truly said to be one of the most serviceable sort growing in fields: and it is my real opinion, that as cole-seed is every year more and more sown, it will, in a great degree, in time, supplant the turnip; because it is most certain, that the colewort or rape will produce more and sweeter milk than the turnip, if given to the beast before it is too old and rank; nor is it so liable to choak a cow or sheep that eats it, as a turnip is. Indeed, as to the hoving quality, I must own, it is rather more apt to do the beast mischief than the turnip-leaf. No green vegetable produces more milk nor fats sheep and bullocks sooner, and that in the severest cold winter or spring-weather, than cole or kale; only particular care must be had to their hoving, which all cattle are very subject to that feed on it. In a word, all thefe inclofed ftubble fields, whofe foils are in good heart, and of the ftiff fort, as foon as harveft is home, or in *September*, may be plowed up and fowed with this excellent feed, as it is done of late in our weftern parts of *Hertfordfhire*, chiefly for feeding our ftore ewes and their lambs in *December*, *January*, and other fucceeding months, when our turnips, and all other field-feeding vegetables are eaten up; by which the lives of multitudes of lambs may be preferved, that otherways would die for want of milk, or other nourifhing food, to enable them to refift the chill of rains, the dampnefs of the earth, and the tempeftuous nights that frequently happen, and prove fatal to numberlefs of thefe new yeaned creatures; for it is common for great rains and fnows to fall in the lambing feafon; and, when thefe happen to be extreme, we fometimes lofe almoft half the number of our lambs, for want of proper food enough to feed the ewes, and breed milk, which this moft fucculent plant will do beyond all others. It is alfo by the help of this cole, kale or rape, that the fuckling of calves may be carried on in the greateft perfection; and that when the farmer is not only deprived of all other field fubfiftence, but where even grains, malt duft, or any thing elfe cannot be got, and this by only the help of hay and ftraw, and this juicy plant. This is one

one example, among many, that discovers the ignorance and bigotry of our *British* farmers, who, though they have often heard of the improvement of field coleworts for more than forty years, yet would never be brought to sow the seed in our *Chiltern* country, till within these very few years past, and that by means of only ocular demonstration of a neighbour's success, which sometimes prevails over their obstinacy, when nothing else can.

These that were sown in *July* or *August*, or *September*, may be fed 'till the beginning of *May*, with sheep, cows, or bulls, and the same land fallowed directly for turnips or wheat. These afford the farmer a vast opportunity of profit, for this, like rye, may be fed in the spring, when grass and green corn cannot; but the cole or rape exceeds the rye, in that it may be fed earlier in snows and near as late, for these will suffer themselves to be eat down two or three times, and quickly recover again, which cannot so well be said of rye. The roots of these, being of the cabbage kind, draw very much nourishment from a rich ground, that forces them on, and gives them a new head in a little time, which again returns the dung and stale of the sheep, to the great enrichment of the land and profit of the owner, who by this means may keep his money at home, which
otherwise

otherwife muft be expended in dung, foot, afhes, horn-fhavings, rags, and other chargeable diftant dreffings, that are not fo good and natural to the land as this of fheep. The months of *July, Auguft, September*, are the propereft times in the whole year for fowing cole-feed.

This moft valuable plant has but a few years paft become common in fome of our fields, and now is known but in few parts under the management I am going to fhew. It is not only profitable for the oil the feed produces, but by confequence it muft be more and more in requeft for improving low, moraffy, and fenny grounds, by feeding cows, fheep, and other creatures with it in hard weather, when no other meat can be had abroad; and it is on this account, that they fow cole-feed about *Weft-Hyde*, near *Rickmanfworth*, and other places, for the ewes that fuckle houfe lambs, who even in fnowy feafons can come at their tall fucculent heads, when turnips, raygrafs, clover, and all other green vegetables are under cover. It breeds a great deal of milk in a little time, both in fheep, and cows, and will bear feeding down more than once. At this place they fow one bufhel of cole-feed on ten acres, which they commonly harrow in on one ploughing, as foon as the forward Puffin pea is carried off the ground. But about *Sax Mundum*, in *Suffolk*, and some other parts, they put the seed to no other use than making oil with it. Another advantage belonging to this serviceable plant, is, that in case your turnips miss taking, you may, if the ground is proper for it, sow it with cole-seed. In our high broad-land loams in *Hertfordshire*, after the last ploughing, we harrow twice or thrice, one way first, then sow the seed, and harrow it only once in a place cross-wise: But in ridge, or two or four bout lands, cross harrowing won't answer. This seed must not be sown too thick where it is not intended to be hoed: But if you do intend to hoe it (which is the better way) then the thicker you sow it, the surer the crop. The roots of this plant, as they draw the earth very much, it forces on, and gives the stalks a new head in a little time, and which, as they are greedily eaten by cattle, is quickly returned again in dung and urine, to the great enrichment of the land; by which the owner is enabled to lay out his money in other necessary incidents, since the great cost of manures is here in a great measure supplied by sheep, cows, &c. But it is not proper to let cows feed on them in the open field, because they are apt to pull them up by their roots; in which case, the way is to cut off their heads, and

Chap. I.   OF RAPE.

and give the cows them in another field, or under covert, and that not in too great abundance, left it hoves and kills them; besides which ill property, when the colewort is got old, it makes the milk rankish. On the contrary, the virtues of rapes or coleworts are, that they are excellent boiling herbs in frosty seasons, quickly fatten cows and sheep, and very opportunely yield a subsistence for cattle after turnips are gone. They cause plenty of milk, and are a very agreeable wholesome food for hogs, rabbits, the great geese, ducks, pheasants, and several other creatures, who greedily feed on their juicy pleasant leaves. If the cle, or kale, or rape, is like to be a full thick crop, it may be fed at times in *January, February,* and *March,* in case the same ground is designed to be sown with wheat at next *Michaelmas,* because in *April* the fallow season should begin for the field to be ploughed the first time; and when the cole grows into such a plentiful crop that it kills the weeds, and hollows and fines the earth, there is a second advantage that may be enjoyed from it. Give the land one or two ploughings, and harrow in turnip-seed: Thus, if the ground is in good heart, it need not be dunged, for then it will do without any manure, and return an early crop of turnips, that may be fed time enough to sow wheat on the same. But if the cole is like to be a scanty thin crop, it is commonly fed off by *Allhallontide*, or the latter end of *November*, in order to prepare the same ground by several ploughings for sowing it with barley the spring following. Where the land is not too wet, some have ventured to turn in and fat bullocks upon the cole heads, with hay constantly by them; but here they are apt to tread them down and spoil many, and sometimes get hoved, to the endangering of their lives. But sheep will fat on them more safely, yet not without hazarding their safety. Of this sort, the suckling ewes are the least exposed to this misfortune. Coleworts will do well on those clayey wettish loams, where turnips must not be allowed to grow, because their broad round roots leave holes behind them, and let in the water to the souring of the ground, and spoiling it for succeeding crops of corn.

In *Scotland*, some think that neither coleworts, turnips, nor carrots, will over-stand the severity of extreme cold winters. In answer to which, I will begin with the objection as to the coleworts. Rapes, or cole, or coleworts are synonymous terms for one and the same vegetable; yet confound the notion of many persons in the judgment of it; for my part, I know of no difference in the species of them.

If

If one ask at a *London* seed-shop for either cole or rape-seed, it is one and the same; so is the herb, whether called cole or rape; which is a very succulent or juicy plant, and therefore more productive of milk in beasts, than either turnips, carrots, or any other sort of root or herb that commonly grow in fields. It is on this account that most of the plough-farmers, for about twenty miles round *London*, who suckle house-lambs, and keep milch-cows, sow this seed; for the high-growing colewort will become an excellent food in frosty and snowy seasons, when turnips, and all kinds of grass, are covered with snow, and produce such quantities of milk, as nourishes and fats their house-lambs with great expedition, for an early and profitable market. But in *Lincolnshire*, and in the isle of *Ely* in *Cambridgeshire*, and in some other parts, they sow rape-seed, in order to obtain full crops of this herb, for feeding and fattening their large poll wether-sheep: which it accordingly does; insomuch that from these places many thousands of cole-fed sheep are driven up to *London* early in the spring; and so early, that by this means they commonly get a better price at this time of year, than in after and later seasons. Hence it is, that our chiltern, and other butchers, keep their fatting sheep back, from sending them to be sold in *Smithfield*;

saying,

saying, Let the fen cole-sheep be sold, and then we should meet with the better market for ours. Not but that we sow great quantities of cole-seed of late, in our chiltern country, in clayey loamy, and even in gravelly loamy grounds, whether they lie dry or wet. But it is past contradiction, that cole thrives better in wet than in dry lands: And, for the better explanation of this, I shall be somewhat particular in writing of the several sorts of soils cole-seed may be sown in to advantage; and first of

Sowing Cole-seed in clayey Ground.

There are several seasons in the year that admit of sowing cole-seed, according to the nature of the land, and the uses it is to be put to. If the land is of the clayey, or very stiff loamy sort, and lies wettish, as most of this sort of soil does, it must certainly require the more ploughings to bring it into a fine tilth; and a fine tilth is indispensably necessary to be prepared and got ready for sowing this seed, whether it be in fen, vale, marsh, or chiltern grounds; for this seed, being about the bigness of turnip-seed, is commonly harrowed in, and not ploughed in; because, if it is done in the last mode, it would be apt to be buried, and the crop lost. But this is not all: The manner of ploughing clayey and stiff soils, preparatory for the reception of this seed, and the growth of its plants, is a matter of importance ; for, without a knowledge of this prime work, the feed, labour and expence, may be loft. In fen, marsh, and vale-lands, they are forced to plough and lay their earth in broad or in narrow ridges, as the ground lies drier or wetter; if very wet, the ridge-lands muft be laid the higher, and fhould be the narrower; and this is to be done with either the common foot-plough, the fwing-plough, the draught-plough, or the foot-turn wreft-plough, &c. generally in the length, and not in the crofs way of ploughing ; for fuch ridge-lands will not admit of any other way of performing it in right order. But where ground lies not too wet nor too dry, and yet is of a ftiff nature, it is in many parts ploughed into four-bout lands, preparatory for harrowing cole-feed in the fame; as I have feen done in fome places in *Suffolk* and *Middlefex*. Now when fuch forts of land are ploughed and dunged well, fo that it lies pretty fine, then the latter end of *June*, or beginning of *July*, the plough-man, who is commonly the fower of this feed, harrows it firft twice in a place ; and then proceeds to fow about three or four pounds, or more of it, on one acre; and then gives it only one harrowing, and the work is done : But fome will fow half a peck on one acre of ground, for the better fecurity of a full crop.

Dunging

Dunging and manuring land, for a crop of cole or rapes, is to be done before or after the cole-feed is fown: If before, dung is a very agreeable dreffing for nourifhing a crop of rapes; if after, foot, or peat-afhes, or clay-afhes, or oil-cake powder, or any other pulverized fertile manure, will very well anfwer the end. If the dung is of the long fort, it is very proper to plow it into the ground in *March* or *April*, that it may have time to rot and mix with the earth, againft the time of fowing the feed in *June*, *July*, or *Auguft*. If it is fhort rotten dung, it is beft ploughed in at the laft ploughing; for fuch fhort dung will mix with the earth at once, and is a very natural dreffing to this fort of plant in particular; becaufe a colewort being of the cabbage-kind, it employs a confiderable quantity of earth to nourifh its growth; and where a good parcel of dung is laid in to its affiftance, this plant will grow very faft into a large head of leaves, and thereby be the better enabled to refift the feverity of a long frofty or wet winter. But notwithftanding dung is fo agreeable to the nourifhment of a crop of coleworts, yet there is an inconveniency attending it; and fuch an one, as oftentimes proves fatal to whole fields of rapes; for dung is well known to be a great breeder of flies and worms, and an incentive to the deftructive flug, or naked fnail;

and

and which are all enemies to a crop of rapes; becaufe while the rapes are in their tender infant growth, and when they have made their firſt two leaves, the flug (eſpecially if wet weather happens in the time) is very likely to attack and eat them up; and then, very probably, the whole field muſt be ploughed again, and another quantity of feed of this, or fome other vegetable, be fown to a farmer's great lofs. I have loſt whole fields of coleworts in a very few days, that were eaten up by the flug, or little naked fnail, while they were in their firſt and infant growth. As no land can be too fat and rank for the growth of coleworts, fo none can over-dung, or drefs it too much, for the growth of this open-headed vegetable; and therefore new-broken up ground, made firſt fine, or any other rich moiſt foil, in chilterns, in fens, or in marſhes, cannot be too rich for bearing a crop of thefe plants; and which gives thofe perfons, who are owners of fuch foft ground, that lies near the falt water, a favourable opportunity to improve it to an high degree of profit, by fowing it with this feed, where it will flouriſh in very large and high ſtalks and leaves.

It is true, that the moſt common way of fowing cole-feed is, out of a man's hand, in the broad-caſt mode of fowing it, without the affiſtance of any matter mix'd with it, to make

it

it spread the better, as Mr. *Worlidge* and others direct; for this seed is bigger-bodied than clover-seed; and if a person were to advise our countrymen to mix sand, or other light earth, with it, for preventing it growing in clusters, they would be apt to laugh at him. I never knew any of this or clover, or lucern, and some other small seeds, sown with any mixture, but intirely of themselves; and most of our ploughmen are expert enough to sow them so true, that the crop generally comes even; and this sort of broad-cast sowing they perform not only on broad lands, but also on four-bout lands, and harrow-in the seed: Yet there is a way to get a crop of rapes by the drill-plough, which will sow the seed in a regular manner in drills, at a foot distance, and cover it with mould, immediately after the seed is dropt out of the hopper; and thus become a dressing for nourishing the crop under all its growth. But this is not all the dressing that a drilled crop of rapes requires; for, besides the mould that always falls in of itself, as soon as the seed is dropt into the earth, there is more to be added afterwards, when the *Dutch* hoe or horse-break is employed in the interspaces between the drills; which leads me to make observations on the same.

Either the *Dutch* hand-hoe or the horse-break may be made use of: If the hoe is to be be employed, the rape-feed fhould be drilled in drills at a foot afunder: If the horfe-break is to be employed, the feed fhould be drilled in drills at a foot and an half, or two feet afunder: But whether the one or the other is employed, there muft be two feveral hoeings, or two feveral breakings of the ground, between the drills, if their growth will admit of it, to kill weeds, and lay fome mould on the roots of the rapes. But I fhould have firft faid, that if the rapes come too thick in the drills, there may as many be hoed up with the common pull-to hand-hoe, in their infant growth, as you think is neceffary for giving thofe rapes that remain behind, and are to ftand for a crop, fufficient room to be maintained in a fertile and large growth: For this new way of improving a crop of coleworts will turn to great advantage, if a confiderable quantity of ground is fown with them, and the climate will permit them to ftand the winter *; becaufe thefe two new inftruments are exceedingly well contrived for the purpofe, not only for difpatching a great deal of work in a little time, but for doing it in fuch a manner as faves the expence of dreffing. What a valuable inftrument then muft a right fort of horfebreak be, when, by its being drawn along the intervals, it not only loofens the ground, and kills

---

* Relative to *Scotland*. EDITOR.

kills the weeds, but at the same time lays such a quantity of mould over the roots of the rapes, that as the autumn rains wash down the salts of such earth on them, they will force on both a very quick, and a very large growth of the coleworts! Here you see how necessary a three wheel drill plough, an horse-break, and a *Dutch* hand-hoe, will be for sowing marsh or moory ground with cole or lucern-seed; I say, necessary, because this piece of husbandry may cause such a soft soil to pay you as much, or more, than the driest ground. But where the drill-plough, the horse-break, and *Dutch* hand-hoe, are unknown, the cole-seed is sown broad-cast; and then some will hoe the crop at a proper height, with the common pull-to hand-hoe, in the same manner they do turnips. Others will not hoe them at all; but when the crop is to stand for seed, it is more than ordinary necessary to have coleworts hoed one way or other.

Chusing right cole-seed is a material article; for without good seed a good crop cannot be depended upon. In *November* 1744, I was commissioned by a gentleman living in a distant country, to buy a bushel of cole-seed. Now, as several places lay very convenient for this purpose, I bought this quantity at one of them; but I first tried more than one, because I was offered a brown, and almost reddish-coloured cole-seed, for four shillings a bushel,

but

but I rejected it as worthless; for such seed was either decayed by long keeping, or housed wet, and heated (or what we call burnt) in the mow; and then it is spoiled. At another place I was offered cole-seed from three to five shillings a bushel, and chose the latter, as being of a jet-black shining colour, with a large heavy round body: This I bought, and am sure, if no ill accident happens to it, it will give the gentleman satisfaction.

Rapes are now well known in several countries to become field, as formerly they were only a garden-ware; and indeed this is done with a great deal of reason, as the rapes serve to breed up young cattle, produce milk in cows, ewes, sows, rabbets, &c. and fat dry beasts with the greatest expedition. Of late, therefore, rape-seed has been sown in our chilterne country with great success; but as this seed is for the most general part sown in the broad-cast form, and only harrowed into the surface of the earth, it is very much exposed to the damage of dry weather, which is an utter enemy to the growth of this succulent plant; because this, like the bean, requires much moisture to perfect their large growth. And hence it is, as I said, that in fenny, marsh, and vale-lands, they run into the biggest size, and prosper to that degree, as to yield abundance of seed.

BOOK XIII.

Of several CROPS *not commonly cultivated.*

Carrots.

A Person of note, in *Gloucestershire*, said he could fatten an ox with carrots as much in one week, as common meat would do in four.

Potatoes.

This root is sometimes cultivated by the help of fern: There was a bricklayer, that lived near *Box-Moor*, who was observed to get the largest and most potatoes of any of his neighbours in his private garden, chiefly by the use of fern; for, as he lived near ground that furnished him with plenty of this weed, he used to lay it in trenches, instead of stable-dung, believing it exceeded dung in its usefulness, by furnishing them all the summer with such a moisture, as supplied a watering-pot; for all fern is of such a spungy nature, as makes it easily retain wets, when once it is thoroughly soaked by them; and so does old thatch.

Canary.

This seed, or grain, growing both in light and stiff lands, is generally ripe in *August*, and in *Kent*, is cut down three several ways; one is by the hook and hinks, a second is by the sickle or the reaping-hook, and a third by the scythe as it stands. When it is dry enough, they take and bind it up in sheaves, or bundles, and after it has had its due sweating in the mow, a man, if it yields well, will thrash out three bushels in a day. If the ground is in good heart, and a kind summer follows, one acre has produced three seam and a half, or what we call three quarters and a half. But on a particular acre there have grown four quarters, that have been sold for ten pound a quarter: but since it has got cheaper, and now generally sells for thirty or forty shillings a quarter. Its seed is an excellent sort for feeding cage-birds, and making one of the whitest and best of oils for the limner's use. Its straw is good for cows, but sheep will not eat it; for horses it is indifferent, and therefore cut it into chaff.

Weld.

A load has been had off one acre, and sometimes a load cannot be got off ten. Twenty-eight bundles make one load, which, in the year 1738, was worth ten guineas; and, near *London*, they carry a load and a half in a waggon at once.

Saffron.

An acre commonly yields ten pounds weight, and sometimes as much more, and sold at dearest for five pounds a pound weight; but at the cheapest for twenty or thirty shillings a pound.

BOOK XIV.

Of the CULTURE of CLOVER.

CHAPTER I.

OF THE TIME OF SOWING.

THIS is one main article in the art of sowing clover; for, if it is sown too early, or too late, it may occasion the loss of the crop: for this reason no prudent farmer will sow clover-seed between *Allhollantide* and the first of *March*; because, if he does, he exposes it to the damage of frosty weather, which oftentimes becomes fatal to the clover-seed sown in that time. Hence it is, that many farmers conclude, that clover is mostly killed by frosts, and not by the slugs; by reason when clover-seed is sown too early, and the frosts meet it on the chip, or first infant sprout, they generally destroy it, especially in case the frosts are any thing violent. I remember

Chap. I. OF SOWING. 485

member that, in *March* 1741, the shop-keepers, who sold clover-seed, said, they hoped clover-seed would grow dearer than three pence *per* pound, as it was then at, because they hoped the forward sown seed, by the frosts that followed, would be destroyed; and on this account it is, that few farmers venture to sow it before *March*, for even then it is not free of this sort of danger. And it likewise may be killed by too late sowing; for, if this seed is sown in *April* and *May*, and a long season of hot and dry weather succeeds, it may dry up the seedling shoot, and spoil the crop, especially where clover-seed is rolled in; therefore the month of *March* is the most proper month of all others for sowing clover-seed; and the more, for the *March* winds are commonly so sharp and cold, as to hinder the appearance of slug, fly, or worm, at the surface of the ground, which gives the growing clover-seed a security of making its first sprout free of their rapine. But it is likewise the most proper month of all others to sow clover-seed in, because barley, oats and pease, are more sown than in any other month; and, as clover-seed, more than any other, is sown among wheat, barley, oats, and pease, it is the best time of all others to sow and propagate this excellent seed in the several manners I am going to shew:

CHAP. II.

OF THE CROPS WITH WHICH IT IS SOWN.

SOWING clover among wheat, in *March*, is put in practice, becaufe all tilth land, fowed with wheat, is firft prepared and brought into a porofity or finenefs for the reception and growth of the golden grain; and that the wheat crop may be a full one, fuch ground is commonly dunged or folded upon, or otherwife dreffed or is manured. When the land by feveral ploughings is thus prepared, and dreffed befides, for fowing of wheat-feed in, it is the fitter to fow with clover-feed in *March*, becaufe it, being in a clean hollow, rich condition, will prefently draw in the clover-feed that is fown broad-caft over it; for fo ready is fuch prepared earth to receive, nourifh, and produce a good full crop of clover among the wheat, that fome do nothing elfe but fow it, and thus let it take its chance for growth, and it oftentimes hits fo well as to grow into a full crop. But others take more pains to get a crop of clover among wheat; and, if they fow this fmall feed among wheat that lies in two-bout ftitches or ridges, they, after fowing the clover-feed broad-caft, to the quantity of about ten or twelve pounds on each acre, employ the great, folid, wooden roll,

Chap. II. OF THE CROPS, &c.

roll, by drawing it acrofs the ridges of the wheat; and by this means they prefs down the feed into the earth, clofe the furface, and very much fecure the roots of the wheat and clover, againft the prejudicial power of too much heat and drought. Others fow clover feed over wheat, when it lies in broad-lands, and likewife roll all in for the fame purpofe. Others that fow clover-feed among wheat that lies in broad-lands, will draw fome bufhes, or hafle, or other rods between the teeth of the harrows; and, as foon as they have fown the clover feed, will by drawing a pair of harrows fo prepared over the wheat, raife earth enough to cover and hale the feed. Others, to anfwer this end, will draw bufhes through a hurdle, and afterwards draw the fame hurdle over the new fown clover-feed. In any of thefe forms cloverfeed feldom ever fails of growing into a full crop among wheat, by reafon the earth is in the richeft condition for a wheat crop; and therefore a farmer may depend on having good fuccefs this way, if accidents of weather do not hinder it. But, before I quit this article, I think myfelf necessarily obliged to take notice of one great inconveniency that attends the fowing clover-feed among wheat; and that is, in cafe a hot fummer is attended with frequent fhowers of rain, the clover and wheat is apt to grow luxurious, to

that

that degree, that the clover being thick and long, shades the wheat-roots so much, and lodges so much water in it, as to cause the large stalks of such wheat, with its heavy green ears, to fall down, and be laid before it is ripe, and then the consequence sometimes is the loss of a valuable crop, in a great degree. However, where clover is sown among wheat that grows in poorish ground, the danger is the less; and therefore I know that many farmers boldly venture to harrow clover-seed in among wheat in the manner I have before mentioned, where their land is of the gravelly, chalky, or sandy sort; I mean, where there is a mixture of loam enough among either of them, to make them be called a gravelly loam, a chalky loam, or a sandy loam; for a naked gravel, chalk, or sand, is an improper soil to sow with clover-seed.

By the greatest farmer with us, clover is seldom sown amongst his barley, because of its rank growth in a wet summer, even to the endangering the miscarriage and crippling of the crop; for that the barley is less able to withstand its luxuriancy, than either the wheat or oat, and that for these reasons—The stalk of barley is generally shorter than either of the other two, as being not so long jointed as the wheat or oat, which is visible in particular at the upper knot of these stalks;

for

for upon examination, the higheſt joint of the wheat or oat is commonly found to be as long again as that of the barley; and therefore they are more out of the clover's power, by ſo much as they are higher, and have a greater potency to ſhade, cover, and retard its growth; this generally induces him to ſow his clover in *February*, on this wheat ſtitch, which being well dreſſed and in a fine tilth, often takes extraordinary well; this he feeds one year intirely, and the firſt part of the next ſummer, till he mows or feeds off the firſt head, when he ploughs it up two or three times, and gets either a crop of turnips, or wheat again; or eats it all the ſecond ſummer, ploughs it up in winter, and ſows it with barley the next ſpring. But of late he only lets it ſtand but one ſummer, as thinking the ground becomes ſour by feeding it two. Some ſow the clover on wheat in *April*.

Another, I knew, that ſowed his wheat ſtitches with clover in the ſpring, and for a trial, got a large bundle of buſhes tied faſt together, and by the end of the rope drew it up and down the ſtitches of half the field; this received the ſeed very well, but the other half, that he did nothing to after he ſowed it, had no effect.

When clover is to be ſown among barley, the management of ſowing it is different from that of sowing clover among wheat; for, after the barley is sown and harrowed in, we sow our clover broad-cast, and harrow it in only once in a place, and it is done withal, till the barley is to be rolled, which commonly requires to be done in about a fortnight or three weeks time. This is the most general way of sowing clover among barley, because the harrow tines hale and cover the clover seed better at one harrowing, than more; for, if two were to be given it, it would be apt to bury this small seed. Others refuse this method of sowing clover, and only sow the clover-seed just before the barley is rolled, in order to keep the clover from being too luxuriant, and by that means getting such dominion over the barley-crop, as to spoil or greatly damage it; which is often the case where clover is sown with the barley, and a wet hot summer succeeds, especially where the clover-seed is sown in a rich soil; which misfortune is prevented by the late sowing of clover-seed till the barley has first got high enough to become its master, and keep its growth in low order all the rest of the summer.

This is a pretty sure way to secure a barley-crop from this damage, because as the clover-seed is sown and only rolled upon the surface, it requires more time to take root, than when it is laid deeper by the harrow-tines; so that by late sowing

it,

it, and only rolling the feed in, the barley-crop is never hurt by the clover; and, indeed, this is a matter that deserves consideration; for, suppose a poor farmer had most of his rent depending upon his barley-crop, and this crop should be half lost, merely by the means of his sowing clover-seed when his barley seed is sown; what a loss must this be? I have known this very case befall a farmer in *Hertfordshire*, who rented a very large farm, and sowed, I believe, thirty acres of barley, mostly with clover-seed, that by the frequent showers of rain that followed the sowing, brought on such a rank growth of the clover-grass, as made it destroy great part of the barley-crop; and, as the farmer was then but weak in pocket, it had like to have broke him.

But this is not altogether the ill consequence of harrowing in clover-seed with the barley-seed. There is another attending it at harvest, that is when the clover-grass has got into a very rank growth, it is apt, after mowing, to retain dews or rains, to that degree, as to hinder the barley from drying; and, if barley is not got in dry, then in course there will a short price attend its sale at market.

To this may be added one more loss that sometimes happens by this very means to a barley crop, as when the clover grows so rank among it

that

that it has acquired large ftalks, and that fuch large ftalks by the bulkinefs of both the grafs and barley-crop cannot be got thorough dry, they are liable to damp the barley in the mow, caufe it to look reddifh, and sometimes mufty withal; and then it very likely may be fitter to give fwine, than make malt of, as has been the cafe of many farmers to my knowledge: and I muft confefs myfelf to have been overtaken in this manner; for where a crop of barley and clover is rank of growth, and a wet harveft accompanies its mowing and getting in, the clover has further added to the misfortune, and has caufed the grain to fprout in the mow; and then, where there is much fown, the lofs is the greater; for when it is in this condition, and barley fells cheap, the malfter in courfe refufes it, and then it often falls to the fhare of fowls, horfes, or cows, or fwine.

Wherefore I have this to propofe, that it would be fafer to fow clover-feed in two different manners among barley, than to fow it all one way; that is to fay, if half a barley-crop had clover harrowed in with its feed, and the other half had clover rolled in when the barley has grown high enough for the purpofe. In either form we commonly fow ten or twelve pounds on each of land; not but that even three or four pounds of this fmall feed are fufficient to furnifh

one

one acre of ground with a plentiful crop of its grafs: but, to allow for accidents, there ought to be no lefs than ten or twelve pounds fown. And to grudge feed, on this account, may prove the old proverb true, *Lofe a fheep for a half-penny worth of tar.*

Barley-feed is affuredly the beft *Lent*-grain to fow clover-feed amongft, becaufe, the ground having been under preparation for the reception of the barley, by feveral ploughings and a proper dreffing, the clover-feed will the more eafily and more certainly take the ground, than when it is fown among oats, or peafe, or beans; becaufe thefe three latter feeds have feldom above one ploughing beftowed on them, and are fown without any previous dreffing, which gives the clover-feed the lefs chance of growing into a full crop of its grafs.

A farmer having a large crop of peafe ready to cut with the long *Hertfordfhire* hook, as they grew promifcuoufly; he went over the whole field, and fowed about twelve pounds of clover-feed on each acre, and, by the hookers treading it in with their feet as they worked along, the ground, which became very hollow by the great cover of the crop of peafe, received the feed fo we.., that, next fummer, there appeared a famous crop of clover; for here the feed is pretty

fafe

safe from both flug and fly at this time of the year, in fuch a trodden piece of corn-land.

About *Watford*, where they pay twenty fhillings *per* acre a year for their ploughed-land, they have fuch a regard to the improvement of this noble grafs, that, after the barley is carried off, which the clover was fowed amongft, they get large quantities of *London* coal-afhes, and lay them in a few large heaps in feveral parts of a great field in *September*; then in *October* they turn in thofe fheep, which a little before were put into their ftubble-fields to fatten, and by this opportunity they continue feeding them 'till their turnips are ready, which compleats them for the butcher; then, about *Chriftmas*, they fow their afhes on the clover, which is of great benefit to it.

This feed a farmer harrows in with his oats, without any pole a-crofs them, as being more fure than rolling it in; becaufe the fly, froft, flug, wet, winds, or fun, have not that power to hurt it, when its enveloped in the earth, as when it is confined only to the bare furface, and nakedly expofed to the feveral devaftations; befides, as there is only one ploughing to the culture of this grain, the four, furly quality of the ground has lefs power to hurt the feed when it is harrowed in with
the

the oats, than when it is only sowed and rolled in.

If a crop of clover is intended to be got with barley, then it must be sown a fortnight, three weeks, or a month after the grain, and only rolled in, as I have before hinted, which generally prevents this damage: But here I publish the several methods, that the reader may make his better choice.

If a crop of clover is designed to be had after peas are got off, then when the peas are harrowed, clover seed should be also harrowed at the same time; but here he runs a risque, for if the peas are a great crop, it is a wonder if the clover is not killed by their cover; but if some of it takes, it may be thickened, by sowing a few pounds on an acre, just before the peas are hooked or mowed, which by the thread of the workmen, will be forced into the hollow earth, and very likely become a good crop.

So also may clover be harrowed or rolled in with rye, either in *August* or the spring, and become a good crop, without damaging this grain, whose stalks growing very high, is the less subject to the clover's fury.

Not but that there is a way to come by a good crop of clover, trefoil, or ray-grass, though the seed is sown on a poor ground, and
that

that is this: If artificial grafs-feeds are fown on a poor ground that is firſt ploughed and made fine, they may take ſo well as to cover almoſt all the land, and grow into a thick crop the following year; but as it is not reaſonable to expect there will be a rank growth of ſuch artificial grafs, without any other help on ſuch poor ground, a full dreſſing or manure of foot, malt-duſt, lime, or coal, peat, or wood-aſhes, or ſome ſuch ſort of aſſiſtance ought to be put on the grafs-crop in *January*, or *February* at fartheſt; and then, if a wettiſh time ſucceeds the ſowing of ſuch manure, there will, in courſe, be a large crop produced of ſuch artificial grafs; as is ſometimes proved, but not very commonly; becauſe farmers find it beſt anſwers their intereſt, when they ſow their artificial grafs-feeds on ground dreſſed on purpoſe for a barley, wheat, or oat-crop, for that one ſuch dreſſing very well anſwers both ends; that is, to produce both a full crop of corn and of grafs, and this for two years together where the land is naturally good; but, if it is naturally poor, then one year will be long enough. Indeed, in ſome parts, they let a crop of one or other, or all together, of theſe artificial graſſes lie three years in all in gravelly loams; but then they take care to manure the whole crop the ſecond year for making it hold out ſo long in good order, and this eſpecially is

done

done on dry grounds, that will bear the feet of cattle without their ftolching it; and where fuch artificial grafs is more than ordinary wanted, as the cafe is with feveral of thofe farmers who live between *Hempfted* and *Watford* in *Hertfordfhire*. But, in this management of artificial grafs-feeds among oats, the oats ought not to be fown quite fo thick as when they are fown alone; becaufe too much cover fometimes kills the fprouting infant grafs, which is the fault of, and what difcourages many farmers from fowing clover and other artificial graffes among peafe, becaufe their great cover deftroys oftentimes more than nourifhes a crop of young grafs; and fo does any other corn-crop in a leffer degree that grows very thick, fo as to choak the clover, trefoil, or ray-grafs, while it is in its *embryo* condition of growth; and therefore four inftead of five bufhels of white oats are enough to be fown on one acre, when any or all thefe graffes are fown among them; and then there will be room enough for corn and grafs to grow and flourifh. The fame care ought to be regarded, when grafs-feeds are fown among black oats; where four bufhels of thefe are to be fown on one acre, there fhould be only three bufhels and a half employed, that the young grafs may not be deftroyed by too clofe a cover of the oats;

and although it may be said, that such artificial grass draws so much of the goodness of the ground, as to rob, in some degree, the oat-crop; yet it may be also said, that the shade and moisture, which an artificial grass yields to the roots of the oats, will compensate for all damage that may happen on this account.——It is not too late to sow clover, sainfoine, trefoil, or ray-grass seed in any part of *April*; for it may be done very well among barley or oats, and where a person thinks fit to venture his grass-seed. It may likewise be done among pease, as the tender late-sown puffin-pea, the blue pea, or the *Essex* roading-pea, &c. There are several ways, as I have remarked before, to sow clover or other artificial grass-seeds; but *April*, in particular, gives *Chiltern* farmers a good opportunity to sow them by the nine feet long wooden roll, by reason this roll is commonly employed then in the rolling of that barley and black oats which were sown in *March*, and for sowing either clover, trefoil, ray-grass, seed, &c. among these grains; the way is, just before the roll is to be used, to sow the grass-seed twice in a place, and then immediately draw the roll all over that, and the barley or oats only once in a place, cross the broad-lands; for we seldom ever draw the

roll

roll twice in a place, unless there be an extraordinary occasion for so doing, as when the ground lies more clotty than ordinary; and, if a favourable wet time happens afterwards in due season, there may an excellent crop of such artificial grass appear, when the barley or oats are mowed off; but, if a long dry time follow, it may dry up the sprouting seed and kill the crop; because this superficial way of sowing grass-seeds very much exposes them to this misfortune. However, there is no way, strictly speaking, that artificial grass-seed sown at any time, or in any shape, is free of danger of being spoiled or damaged: If it is harrowed in with corn in *February* or *March*, the frosts may overtake it and ruin its growth; or, if it is bush-harrowed in, its fate may be the same; or whether harrowed in, or rolled in, the fly, or slug, may devour it, before it has got strength enough to resist their rapine. However, the sowing of clover, trefoil, or ray-grass, by the roll, in *May*, is the last resource; and, therefore, all those, who would enjoy a crop of any or all of these most serviceable artificial grasses ought to sow their seed, at farthest, in *April* by roll; which indeed is the best time of all others, to sow any of these seeds among barley in particular, where a farmer has reason to fear, that, by the rankness or richness of his ground, their grafs would grow fo luxuriant as to choak or cripple the barley-crop; as the cafe has often happened to many, when a hot wet fummer fucceeds the fowing of it in a rich foil; but, by this way of rolling the grafs-feed in at this time of the year, he is intirely free from this fort of damage, becaufe the barley-feed, being harrowed in fo long before, has the opportunity of acquiring fo forward a growth, as to be able to keep the clover under, and prevent any fuch misfortune.——Sainfoine feed alfo may be fowed in *April* among barley or white oats, though it is moft commonly done in *March*. If it is to be fown among oats, the fame ground fhould be duly prepared for it, by two or three winter's ploughings to get it into a fine tilth; for, unlefs this piece of forward hufbandry is well obferved, there is no great likelihood of obtaining a full crop of fainfoine, becaufe this is a large feed, that cannot be well covered, unlefs the earth is in a fine hollow condition, for admitting the harrow-tines to enter deep enough for the purpofe; yet, notwithftanding fuch precaution is duly obferved of preparing the ground by feveral ploughings, by the better fort of farmers, there are others of the worfe fort, who carelefly venture to fow this or other grafs-feeds on rough tilths, and mifs their crop by fo doing.

A certain

A certain farmer, who rented a large farm, in *Hempstead* parish, and had no common, fit to feed sheep on, is afraid to sow oats in those of his inclosed fields, whose soils are loamy clays, because he is of opinion, they draw the heart of the ground out so much, that, if a crop of wheat next succeeds them, it will be a poor one. He, therefore, generally contrives to have a crop of clover and trefoil grow together in a mixture, for feeding and fattening his grass-lambs, believing, that, after such a crop of artificial grasses has been fed off, with ewes and their lambs, he shall have a good crop of wheat after it.—The main root of clover is a tap-root, and all the tap-roots love to follow an horizontal or spreading root, as all corn roots are; because each sort gets its living in the ground, in different strata's of earth; yet, I dare not say that artificial grass does not hurt corn, or corn that, when they grow together, as Mr. *Bradley* has asserted; no, for grass may hurt corn, and corn grass, as in the cases of barley and pease. But this I have to say, that all tap-roots have the advantage of being much nourished by the washings of the surface that by long and great rains are forced down to them, which is one main reason why clover yields the farmer two or three mowings in one summer. Another author remarks,

remarks, that clover is very apt to fail in a dry summer: This may be, and so may all other vegetables in some degree; but it is well known that clover has grown into great crops, when natural grass has been burnt up, and this not only from the benefit of its tap-roots, but also from the assistance of coal-ashes, or soot; which, if sown on a first or second year's crop of clover, it will bring the clover under such a fertile forward growth, as to secure it very much against droughts, though no rain fall from *Lady-day* to *Mid-may*. But this last author, I found, in my travels, lived in a country, where they were no better acquainted with wood-soot, than to sell it only to dyers; and, for coal-soot, I suppose, they would think nine pence thrown away, if they were to give a shilling for a bushel of it. Whereas the *Hertfordshire* farmers, even those who live thirty miles distant from *London*, think it good husbandry, to give ten, eleven, and twelve pence for a single bushel of coal-soot, to lay on clover, sainfoine, wheat, barley, natural grass, *&c.* and yet are sure they have not all neat soot brought them for so great a price.—If dung is laid over a second year's crop of clover, it ought to be done in a very thin manner; for, if it is laid in clots, it will be apt to smother part of it, and then, if the clover is mowed, it will rake up with the hay. But this fault would prove the greater, if dung is thus laid on new-sown clover (as is too often done) for then such young clover would be the sooner smothered; but, if stable or yard-dung is laid over any clover, in a light thin manner, it will greatly nourish it.

CHAP. III.

OE THE GROWTH OF CLOVER.

IT commonly comes up in a week's time, if the weather is favourable, and then it appears with two leaves and its seed on its head; these all rot away in about three weeks time, and then the spear, which shot from the middle between the two first leaves, opens itself into three new leaves, which stand all good. Old seed is longer taking root, and its leaves bitterer than those from new seed,

CHAP. IV.

OF MAKING CLOVER HAY.

THE latter end of *May*, or the beginning of *June*; clover is fit for mowing, and known by its being full knotted and red-headed;

headed; and it is then you should begin this work; for there is a crisis of time to be observed in this, as well as for natural grasses. If you mow it too soon, it will shrink and lose in quantity; and, if it stands too long, you will be deprived of its best quality, the sap; and then consequently it will be very coarse, and want much of its due value. My way of making it is thus: After it has lain a day or two in the swarths as it was mown, the next time I remove its first situation, by turning the swarths, and letting it lie so another day; then I put it in grass cocks in rows, that the ground between them may be regularly raked; then I afterwards turn them with a fork topsy-turvy twice a day, 'till the hay is intirely made. By this method you will have it in its due perfection of colour and sweetness with the leaves on; but you must not expect this latter benefit, if you make this hay after the common way of throwing and spreading it several times, as is done in making the natural sort; because then it would be deprived of its leafy and second best principal part.

In the next place, when you inn it, and it should happen to be wet weather, or that you mistrust a dampness in it, put a tub, basket, or hollow square made of four narrow boards four or six feet long nailed together, and placed in

the

the middle of the cock, or stack of hay abroad, or in the mow in the barn, pulling it up higher as the cock, stack, or mow fills in; and thus you will not only prevent its firing, but keep it sweet; and indeed it is a good way to use this method always, let the hay be ever so dry, because it gives an evacuation to all moisture, and tends very much to the preservation of its green colour. If you intend the second crop for seed, do not feed it after the first mowing.

CHAP. V.
OF THE APPLICATION OF THE CROP.

A CROP of clover, which may be, and often is, of more worth to a farmer, than either a wheat, barley or oat-crop; which I pretend to make appear, by reckoning what a full crop of clover-grafs was worth in the year 1743, and a crop of wheat, or a crop of barley, or oats, or peafe was worth at that time, Wheat was sold at *Michaelmas* for less than three shillings a bushel; that which would fetch fourteen shillings a load, as we call *five bushel* in *Hertfordshire*, must be the very best of wheat; and, I will suppose, that there were gotten off each acre of land thirty bushels one acre with another:

another: Then the amount of such an acre of wheat would be at that price four pounds four shillings. But, after ploughings, dressings, and rent, with other incident charges, are deducted out of that sum, there will hardly be twenty shillings left for the farmer. Whereas a full crop of clover-grass will produce three loads of hay, and, as each load is worth twenty-five shillings, the whole first crop will amount to three pounds fifteen shillings, besides a second and third crop of the same grass the same summer, if it happens to be a hot rainy one. But I will reckon no more than the money for the first crop, the last one or two will be more than sufficient to defray rent and all other charges of this clover-crop; and then there will remain clear, into the farmer's pocket, the sum of three pounds fifteen shillings for one acre of clover-grass got off the ground in a fallow-season this year 1743, which was one of the most propitious summers, that ever I knew, for causing the biggest crops of clover; for I had, it was judged, as much, or very near as much clover-hay at a second mowing, as I had at the first mowing: Five or six loads in all off one acre at twice, and so in proportion for a crop of clover after other grain, for this clover-crop, here mentioned, followed an oat-crop that the seed was sown among.

Clover,

Clover-grafs, in full bite, produces a great deal of milk; infomuch, that an acre of this, well planted, is reckoned to feed as many cows as two or three of meadow; but this has its faults, for, at beft, it makes but a coarfe butter and cheefe of a difagreeable tafte, and worfe if it ftands too long before it is fed; therefore it is beft fown with ray-grafs and trefoil, which will make a better butter, and be a safer feed than clover alone.

All *July*, 1739, a neighbour of mine drove his hogs a mile diftant from his houfe, every day, into his clover field, under the conftant care of a boy, who, about twelve of the clock, had them to a pond that lay a pretty way off the field, as having no water there, on which with a fupper of wafh and grains, they throve and grew a-pace.

Some fay, that if a farmer does not fow clover, he cannot pay his rent, as times now go in the farming bufinefs; meaning that as many large downs in the weft, and commons elfewhere, are of late ploughed up, and converted into arable land, and by the new improvement of hufbandry in many places carried on with great fuccefs, grain is now become very plenty and cheap, and like to continue fo. The fowing of clover, and other artificial graffes, is abfolutely neceffary to be done, where the foil and conveniency will admit

admit of it, that a farmer may be able thereby to feed his cattle with it, and save his natural hay to sell, and his oats, pease and beans, in a great degree, and do his land a vast service, and all this for a trifling expence. For, first, as ten pound weight of this seed sows one acre, and that at a common cost, amounts but to three shillings and four pence, at a groat a pound, there is hardly any farmer that cannot afford to sow his land with such cheap seed. Secondly, he may, by this means, employ his fallow ground, which otherwise would lie idle; so that a crop of clover, got here, may be justly reckoned almost a clear profit. Thirdly, Such a full crop of clover will last two years very well, if encouraged once in that time, with the assistance of dung, soot, or ashes. Fourthly, This rare grass, when it covers all the ground, never fails of bettering and improving the land it grows on to that degree, that it is in some parts, where I have travelled, called the mother of corn, because it kills weeds, prevents exhalations, hollows the earth, and leaves so many large long roots behind it, as to become a sort of dressing to it. With the help of clover grass, the suckling of house-lambs is carried on most part of the year within twenty miles of *London*, in summer, by feeding ewes with it in the field; in winter, by feeding them under cover with its hay. With

this

this grafs, and its hay, our plough and cart horfes are fubfifted the greateft part of the year, to the faving of our natural hay and corn; for clover alone, when fed by them in the field, will fupport them under their work, and the fame in the ftable, when fed by its hay out of the rack. Our wether fheep and grafs-lambs we fat with it in the field, and our ftore-fheep are fupported moft part of the fummer by its feed, and thereby enabled to drefs their ground with their dung and urine in the fold, to the faving of many pounds to a large farmer, who muft otherwife be at a great expence to buy manure to fupply this fold dreffing. With clover feveral ingenious *Chiltern* farmers, whofe wives, or maidfervants, are capable of managing a dairy of this kind, carry on the bufinefs of making butter and cheefe from this grafs, which formerly was thought impoffible, becaufe of its hoving quality, and the rank tafte it ufed to give the butter and cheefe made of it. But now feveral farmers have both butter and cheefe made of it in that perfection, that many buyers of them cannot diftinguifh either from that butter and cheefe made from fome forts of natural grafs.

But befides the great conveniency of enjoying clover crops, before mentioned, I have further to obferve, that, with this grafs and its hay, there are many oxen, cows, and calves, fatted: for this

this artificial grafs is of a very fattening nature; infomuch that where a full quantity of it is enjoyed by thofe dry beafts, that are to be fatted by it, it will do it compleatly, with the help of very little elfe: I mean, after they have been fed all, or great part of the fummer; in clover, in winter its hay, and a few oats now and then, will effectually fat any of the large horned kind. Clover alfo, either green or dry, will caufe abundance of milk in cows; and where they can be fed in the day time on clover in the field, and in the night time on natural grafs, as is my ufual way, the milk will be very fweet, and ferve for any manner of ufe. And I do further aver, that where a perfon keeps a right fort of hogs for this purpofe, a field of clover will prove almoft an intire fubfiftence for them the greateft part of the fummer, provided the ftore-hogs can enjoy it in frefh parcels; for if they, or any other beaft, are kept fo long on clover, as to ftain it much by their dung, ftale, and feet, it will have a contrary effect, even the caufing them to pine away, inftead of fatting; which leads me to obferve further of clover-grafs, by the want of it in vale-lands.

This artificial grafs, above all others, is the *Chiltern* farmer's friend, and the vale farmer's enemy. The vale-grafs turf is certainly the richeft turf of all others, becaufe its grafs and

Chap. V.   OF THE CROP.

hay, for the most part, will alone fat an ox or a horse; when that in the *Chiltern* country will not, without the help of corn besides; and therefore before clover got footing into *England*, which it did about eighty years ago, the grass farms of vales were the best farms in the kingdom, because they lett for the most rent of all others. But since clover seed has of late been sown by most or all *Chiltern* farmers, the value of such vale-grass farms has declined; and the more, by reason where their vale ridge-lands are of a stiffish soil, and low watery situation, as most of them are, they will not answer being sown with clover.

An author says, that horses should be staked to a certain length, to prevent the mischief that might otherwise accrue, from the rank feeding of the clover. This is a monstrous mistake indeed, for no man, I believe, ever knew a horse suffer this way; it is true, we sometimes bleed them two or three days after they are turned in, to prevent a plethory, and so we do sometimes, if in natural grass: but if ever he had been an owner of cows or sheep, fed in clover, he would have fixed some such caution on their behalf; because it is too often known, that numbers of them have died; not only at their first being turned in, but after some days, and even weeks have past; though it is certain, the greatest danger is at first, by their voracious greedy feeding, and

the

the wind contained in this frim, rank grass, that in half an hour has hoved their bodies like a bladder; and if not directly run about or stabbed in the flank they surely die.

Clover, by some is disputed, whether it does most service to the ground, by being eat all the summer, or mowed twice, once for hay, and the next time for seed. In my opinion, it advantageth the earth most, when the first crop is mowed, and all the rest eat; for by letting it stand till it is fit to mow, it kills the weed by the cover of its thick, high head, hollows the ground, shades its own roots, and thereby prepares the ground for the better growth of the after-grass. But if fed first, then the roots are exposed to the summer droughts for want of that cover, and consequently give the weed the greater opportunity of damaging the clover.

Many farmers, in our *Chiltern* inclosed country, contrive to sow clover among oats, on purpose to employ and improve the same ground the next fallow season, which otherwise would lie idle. By this they enjoy their land every year, and yet improve it, without breaking through the common covenant in leases, that the tenant shall not cross-crop; for, though the clover may be called a crop, it may be also called a dressing of the ground, because a good crop of clover so enriches it, either fed or mowed, that in some countries

Chap. V.   OF THE CROP.

countries where I have travelled, it is under the appellation of the mother of corn; and, that clover, or ray-grafs, or trefoil, or all together, may take furely, our farmers give their wheat-ftitches two or three ploughings. The firft in *October*, or at the beginning of *November*, in a fhape of ploughing adequate to the nature of the foil; which is an excellent piece of forecaft, or hufbandry, if practifed in ftiffifh loams, as well to kill the twitch, or couch-grafs, as the honey-fuckle, and other weeds, and reduce the earth into a fine tilth for fowing and caufing artificial grafs-feeds to take the ground, and flourifh afterwards. And let me tell you, where fuch land is well planted with clover-grafs, and the clover fuffered to ftand for a mowing crop till the latter end of *May*, and, if it proves a full crop, there may be expected as much profit from it as from a crop of wheat. A proof of this, by many, was experienced in the year 1742, which though the fummer was fuch a dry one, as returned us at *Gaddefden*, and a thoufand other places, not half a full crop of natural-grafs, yet the clover rallied, (as we call it) and, by virtue of a good fhower of rain or two, it got a covered head, enough to fhade its roots, and grew into fuch fine crops, as paid many as well or better than wheat-crops. An example of this I fhall here ftate as follows, *viz.*

Vol. I.   L l   A three

	l. s. d.
A three acre and a half inclofed field, at *Gaddefden*, yielded, at the firft mowing, feven loads of clover-hay, fold for thirty fhillings a load out of the field.	10 10 0
A fecond mowing returned four loads at the fame price.	6 00 0
The after-meath grafs, which was fed by cattle, was worth ten fhillings *per* acre.	1 15 0
Total profit in the fummer, 1742—	18 05 0

CHAP. VI.
OF THE SEED.

How to know good from bad Clover feed.

THE purple fort of this is that which had its due maturation in the field, and an efcape from the heat and burning of a damp mow, and is therefore the right true feed which ought always to be coveted and fown by thofe who hope for fuccefsful returns from fowing it: this is truly that part which is beft of all the three forts, and is the medium of the two extremes. The white or green fort is the unripe part, and miffed of that benefit in the field, which the purple had; for in the ear or head of

this

Chap. VI. OF THE SEED.

this grafs, as well as in wheat and other grains, this seed has its several aspects and proportions of bigness, before it is cut down, and will shew their differences accordingly after thrashing and cleaning: where then this white or green sort is in a large quantity, it is to be rejected; for though such seed may take root and grow, it will prove diminutive grafs, and be sooner overcome by the frosts and wets, or droughts, worms, &c. than the fine purple sort. The reddish sort is the worst of all; though this might be as good as the best, when it was brought out of the field, but was afterwards too much heated or burned in the mow, which occasions its reddish colour, and destroys in a great measure the vegetative part of the seed. From whence I conclude that, where clover-feed abounds most with the large purple sort, it is then so much the more valuable: but, for a farther proof of the goodness of this seed, heat a shovel half red hot, and put some seed into it, the good will snap, and the bad will burn away. Kiln-dried clover-feed may be discovered by its igneous smell.

Quantity to be Sown.

Twelve pound of clover-feed sows an acre of ground well. Trefoil in hull, two bushels sows an acre. Ray-grafs the same quantity.

An author says, that five or six pound of this seed is enough to sow among grain: it is true, that there has been good crops followed this quantity, where it has met with a right ground, management and season; but that happiness is not always to be trusted to; and therefore we generally sow ten or twelve pounds on an acre for fear of the worst, that often happens from the slug, fly, frost, wets and winds: now here is wanting that necessary caution that should have been tacked to it, that the clover must not be sowed at the same time the grain is, but only rolled in, about a fortnight or three weeks after; for if it is, and a showery summer succeed, it is two to one odds, in my opinion, if the crop is not a great part spoiled by the clover's luxuriancy, as I have known it several times do.

END *of the* FIRST VOLUME.

www.ingramcontent.com/pod-product-compliance
Lightning Source LLC
Chambersburg PA
CBHW031946290426
44108CB00011B/699